A HISTORY OF NORTH CAROLINA WINES

A HISTORY OF NORTH CAROLINA WINES

From Scuppernong to Syrah

Alexia Jones Helsley

The History PRESS

Published by The History Press
Charleston, SC 29403
www.historypress.net

Front cover: *top*: Cabernet grapes, Shelton Vineyard. *George A. Denka, photographer*; *bottom*: Rockhouse Vineyards. *Jacob Helsley, photographer, courtesy of Lee Griffin.*
Back cover: *left*: Muscadine grapes. *Courtesy of North Carolina Grape Council*; *right*: Rockhouse Vineyards. *Jacob Helsley, photographer, courtesy of Lee Griffin.*

First published 2010
Manufactured in the United States

ISBN 978.1.59629.952.8

Library of Congress Cataloging-in-Publication Data
Helsley, Alexia Jones.
A history of North Carolina wine : from Scuppernong to Syrah / Alexia Jones Helsley.
p. cm.
Includes bibliographical references and index.
ISBN 978-1-59629-952-8
1. Wine and wine making--North Carolina--History. 2. Wine and wine making--North Carolina--History--Pictorial works. 3. Grapes--North Carolina--History. 4. Viticulture--North Carolina--History. 5. Scuppernong--North Carolina--History. 6. Syrah (Wine)--North Carolina--History. 7. Vineyards--North Carolina--History. I. Title.
TP557.H455 2010
641.2'209756--dc22
2010021179

For George A. and Martha Jones Denka

"When God, your God, ushers you into the land he promised through your ancestors...you're going to walk into large, bustling cities you didn't build, well-furnished houses you didn't buy, come upon wells you didn't dig, vineyards and olive orchards you didn't plant."

—Deuteronomy 6:10–11 (*The Message*)

Contents

Foreword

Two Vineyard Visions

Vitis Rotundifolia

A case could be made that the muscadine—North Carolina's native grape—should replace cotton in the first line of "Dixie." *Vitis rotundifolia*, commonly known as muscadine, is North Carolina's original grape. The scuppernong, one variety of muscadine, is the official fruit of the state of North Carolina. It is the oldest commercially produced grape in the United States, and at one time, scuppernong grapes and wines were the bestselling grapes and wines in our nation.

Duplin Winery is proud to produce fine wine from muscadine grapes, such as Carlos, scuppernong and magnolia. A family business, the winery produced its first wine in 1975 and sold its first wine in 1976. When Duplin opened, there had been two other eastern North Carolina wineries, Onslow Wine Cellars and Deerfield Winery. My brother and I began to grow grapes as a sideline in the 1970s. An out-of-state concern was offering lucrative contracts to farmers who would grow muscadine grapes. By the time the planted vines matured and the grape harvest was ready, supply had exceeded demand. The price offered per ton of grapes was not sufficient to cover the cost of growing them. So, to provide a market for eastern North Carolina muscadine grapes, my brother, Dan, and I opened Duplin Winery. It has been a difficult journey, but we have succeeded, and in 2008, Duplin had fifty-five employees and two thousand acres of grapes.

In my opinion, the muscadine grape makes the best sweet wine in the world and is by far the healthiest (it contains the highest level of antioxidants) of all wines produced. What makes this book so important is that few know and appreciate the muscadine's importance in southern agricultural history. Muscadines have played and can play a vital role in enhancing the economy and lives of those who work in North Carolina and of those who enjoy a delicious glass of muscadine with their evening meal. As I have said on many occasions, "A glass of muscadine in moderation is healthy. In excess, it's deadly. In abstinence, it's a crying shame."

The future of muscadine agriculture may be to provide the best source of antioxidants for the nutraceutical industry. Like our immune system, which produces more antibodies when under attack, muscadine grapes produce more disease-fighting phenol compounds (antioxidants) than any other fruit. This characteristic of the muscadine may be a result of its adaptation to the hot, humid and disease-prone climate of the southeastern United States. The same phenolic compounds that keep the muscadine healthy just may keep you and me healthy as well.

Therefore, as the mottos of the Brotherhood of the Knights of the Vine affirm: "*Per Vitem! ad Vitam!*" (Through the Vine! Toward Life!)

David G. Fussell Sr.
Winemaker Emeritus
Duplin Winery, the oldest active winery in North Carolina, the largest winery in the South and the biggest muscadine winery in the world
Rose Hill, North Carolina

Vitis Vinifera

From the Appalachian Highlands and Piedmont to the Atlantic, North Carolina is a rich region of cultures, climates and resources. As our state becomes increasingly well known for wine production, Biltmore® is proud to be part of the story.

Our roots begin in 1895 with founder George Washington Vanderbilt and his magnificent estate in Asheville, North Carolina. Biltmore became synonymous with the highest standards of hospitality, luxury and attention to the smallest detail, including Vanderbilt's practice of collecting fine vintages to share with guests.

By 1960, George Vanderbilt's grandson, William A.V. Cecil, was managing the estate. Over the next few years as he explored revenue-generating possibilities to preserve Biltmore®, he found himself asking, "What's more appropriate for a French chateau than vineyards and a winery?"

At that time, little was known about commercial wine production in North Carolina, so Cecil turned to agricultural experts at North Carolina State University, who suggested planting native muscadine vines on the estate. But Cecil was not satisfied with the results. Experts at Cornell University then advised trying French-American hybrid grapes; they were disappointing, too. Even the American viticulture pioneers at the University of California at Davis warned that while *vinifera* cultivation might be possible in Western North Carolina, it would be expensive and labor-intensive.

Cecil continued to pursue his dream, hiring sixth-generation French winemaster Philippe Jourdain, whose extensive knowledge of the entire wine business would prove key to the project's success. The first *vinifera* grapes were introduced to Biltmore® in 1978, and the Biltmore® Estate Winery opened in 1985.

From Cecil's and Jourdain's tireless efforts to develop vineyards and produce quality wines to our ongoing research, the care of our vineyards, the craftsmanship of our winemakers and our status as America's most-visited winery, Biltmore® continues to be a leader in the North Carolina wine industry today.

Jerry Douglas
Senior Vice-president of Sales and Marketing
The Biltmore Company

Acknowledgements

The author gratefully acknowledges the assistance of Helen Wykle and Sallie Klipp at the D. Hidden Ramsey Library, University of North Carolina–Asheville; Jesse R. Langford and Kim Cumber of the North Carolina Archives; Donna Kelly of the Historical Publications Section, North Carolina Department of Cultural Resources; Brigitte Smith of the Gregg-Graniteville Library, University of South Carolina–Aiken; Garland Goodwin and Helen Woodward of the Polk County Historical Association; Justin Furr of the North Carolina Wine & Grape Council; Jerry Douglas, senior vice-president, and Chris Barnett of Biltmore Estate Winery; David G. Fussell Sr., founder of Duplin Winery; and especially George A. Denka, president of Shelton Vineyards. The author, her husband and granddaughter spent a memorable day with David Fussell at historic Duplin Winery in Rose Hill. He graciously shared his encyclopedic knowledge of the North Carolina wine industry and his rich and historically significant collection of materials that document that history. In addition, Jerry Douglas and Chris Barnett shared their time and valuable documentation about the Biltmore Estate Winery. Special thanks to Jacob and Terry Helsley, photographers, and Lee Griffin of Rockhouse Winery. North Carolina is fortunate to have such valuable historical resources and fascinating wineries.

Author and readers owe a great debt to my indefatigable indexer, Terry Helsley. As others have noted, a book without an index is like a country without a map or a ship without a rudder.

Prologue

Taste and Eat

And wine gives sparkle to life.
—*Ecclesiastes 10:19 (*The Message*)*

This book is a journey through the long and exciting history of North Carolina grapes and vines. The state of North Carolina has been blessed with native grapes that grew with such wild abandon that early explorers were uniformly impressed. Wine production, however, is another story—one with peaks and valleys and switchbacks. Therefore, *A History of North Carolina Wine* is not a guide to the many award-winning North Carolina wines nor a comprehensive look at all the vineyards in North Carolina today. Rather, it is a tribute to man's long love affair with grapes and their wine. As Edward Antill penned in 1771:

> *Its* [the grape] *juices, when express'd and rightly fermented and purified according to art, partake of a noble spirit truly homogeneous and fit for the use of Man. They gladden his heart, remove to a distance his troubles and cares, cause him to forget his poverty and low estate, and raise him to a level with the rich and great: They enliven his thoughts, exhilarate his spirits, cheer his soul, and for a time make him as happy as his present condition is capable of.*

This book recounts the story of North Carolina's volatile wine industry—the story of the promise that was long unfulfilled, of disappointments and success and of competing visions and grapes. This book is for the reader intrigued by the romance of the native muscadines, appreciative of the complex varieties of North Carolina wine and fascinated by the enduring drama of human beings and their dreams. In the Old North State, the highly acclaimed vineyards of today have deep roots in the state's past.

Chapter 1

The Land, the Grape and the Dream

The like abundance is not to be found.
—Arthur Barlowe

In the beginning were the grapes and, in particular, the great white grape: the scuppernong, North Carolina's state fruit. English, French and Italian explorers found native grapes growing in great profusion. These were the native muscadines (*Vitis rotundifolia*) and fox grapes (*Vitis labrusca*). Of interest, these two species are the native grapes most used in modern wine production. The bounteous supply and size of these large grapes impressed early explorers such as Giovanni da Verrazano, Philip Amadas and Arthur Barlowe and the first settlers. In 1524, the Italian Verrazano, while exploring the coast of North Carolina for King Francis I of France, noted "many vines growing naturally." He compared them to the grapes of Lombardy in his home country of Italy and added that "without all doubt, they would yield excellent wines." In 1584, Barlowe, who with Amadas explored for Sir Walter Raleigh, wrote in *The First Voyage to Roanoke*:

> *We viewed the land about us, being, whereas we first landed, very sandie and low towards the waters side, but so full of grapes, as the very beating and surge of the Sea overflowed them, of which we found such plenty, as well there as in all places else, both on the sand and on the*

> *greene soile on the hils, as in the plaines, as well on every little shrubbe, as also climbing towards the tops of high Cedars, that I thinke in all the world the like abundance is not to be found; and my selfe having seene those parts of Europe that most abound, find such difference as were incredible to be written.*

Other explorers of the southern Carolina coast echoed those sentiments. For example, in 1663, William Hilton noted that "the Country abounds with Grapes," and Robert Sandford in 1666 wrote of "diverse grapes vines which though growing without Culture in the very throng of weedes and bushes were yet filled with bunches of grapes to admiration."

Such reports of "abundance" and grapes growing profusely along the shore, among the shrubs, on the plains and draping the tall trees impressed the financial backers in England. In 1606, John Smith also commented on the presence of wild grape–bearing vines near Native American villages. The vines produced better fruit in the sunlit clearings around Indian villages than climbing in the forests. The Indians of the southeastern United States gathered *Vitis labrusca* and *Vitis rotundifolia* for food in the fall. These native grapes were eaten fresh or dried to supplement winter diets. From the native *Vitis labrusca*, colonists developed the Concord, Catawba and Isabella grape species. The scuppernong, however, was the most popular of the *Vitis rotundifolia.* According to ethnobotantist Earl Core, colonists derived the name scuppernong from the native Algonquian word *askuponong*, meaning "place of the sweet bay tree" (*ascopo, Magnolia virginiana*). Scuppernong is also the name of a river in Tyrell County in eastern North Carolina.

Indeed, in 1670, one of the goals of the eight Lords Proprietors set for the new colony of Carolina (which at that time included both North and South Carolina) was producing wine for export. Grapevines were among the cuttings and seeds sent with Joseph West for experimental planting in Charles Towne, the capital of the new colony. In January 1671, Joseph Dalton, a member of the colony's first council, wrote Anthony, Lord Ashley, one of the eight Lords Proprietors, a report on Carolina. Dalton asked for grapevines, as he considered them one of the "profitable communities" available to the fledging colony. He also noted: "Wee have indeed plenty of diverse sorts of grapes here some very pleasant and

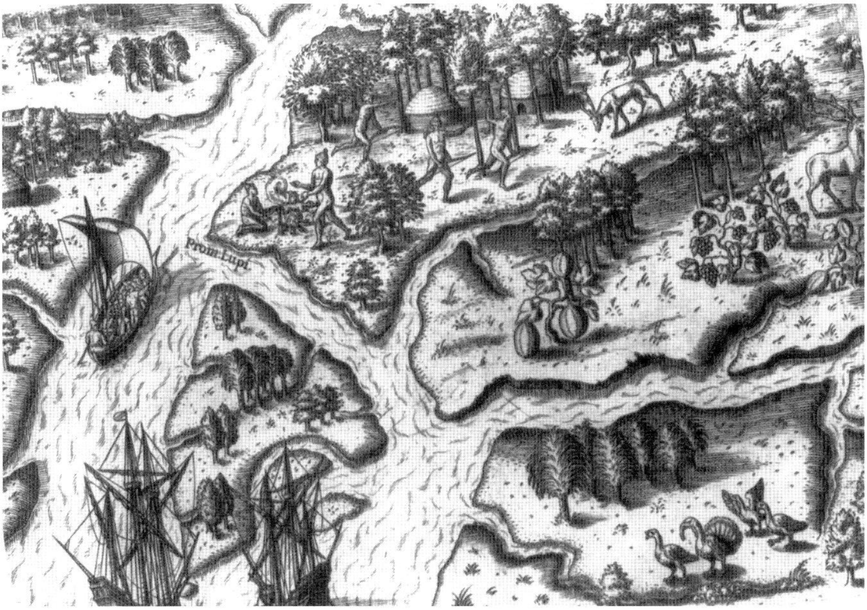

Caddo Indians gathering grapes. Image by Theodore de Bry (1528–1598). Centuries before Europeans first saw the coast of North America, Native Americans supplemented their diets with native grapes such as the muscadine (*Vitis rotundifolia*) or fox grapes (*Vitis labrusca*). *Courtesy of the Library of Congress.*

large but being prest the thickness of their outward skinn yields a kind of harshness which gives us reason to feare (though we intend to make a tryall of them) that they will hardly ever be reclaimed."

Despite Dalton's fears, muscadine grapes did become a colonial dietary staple. The same could not be said for the relations between the northern and southern parts of the colony of Carolina. Communication and transportation were difficult. Indeed, governing the vast expanse of North Carolina from Charles Towne was not practical. So, while North and South Carolina initially shared a government, time and distance worked to separate them. After a number of years, in 1712 the colony of North Carolina had a separate governor.

Consequently, from the beginning, European explorers, developers and settlers had contemplated the commercial possibilities of North Carolina grapes. In addition, the early settlers experimented with the cultivation of European varieties. Yet despite the abundance of grapes and the efforts of promoters such as John Lawson, Gabriel Johnston and

Arthur Dabbs, winemaking did not become a major industry in colonial North Carolina. But the dream persisted.

John Lawson (1674–1711) was an adventurous young Englishman from Yorkshire who sailed for Charles Town in 1700 to see the wonders of the New World. Arriving in Charles Town on August 15, 1700, Lawson and several European and Indian companions made a memorable trek across the Carolinas. In his journal, Lawson recounted flocks of birds so thick that they blocked the sun, trees so tall their tops were obscured and other natural observations. Lawson also reported on and documented Native American life and foodways, as well as indigenous flora and fauna. After his trailblazing trip through the interior of South Carolina, the intrepid Lawson looped back and surfaced near Bath, North Carolina's first town. His journey of exploration covered 550 miles. Besides observing, Lawson also collected specimens for James Petiver (1663–1718), a London apothecary and fellow of the Royal Society. Some of Lawson's specimens are still part of the Sloane collection of the British Museum.

In eastern North Carolina, Lawson settled on Lawson Creek near the site of New Bern. He became an officeholder, returned to London and completed the revision of his journal, continued his explorations into Virginia and became a successful land surveyor and Surveyor-General of North Carolina. Beginning in 1709, Lawson published his journal in London. The journal appeared under various titles, including *A New Voyage to Carolina; Containing the Exact Description and Natural History of that Country: Together with the Present State thereof* and *A Journal of Thousand miles travel'd thro' several Nations of Indians*. In addition to English, Lawson's account also appeared in German and has been reprinted many times. It is a valuable resource for the history of the colony during those critical early years of contact.

In addition to his other pursuits and responsibilities, Lawson retained his interest in the natural world, especially the flora of the New World. As part of that interest, he planted red and white European grapes and studied the varieties of *rotundifolia*. His goal was to develop a viable vineyard and produce wine. However, in 1711 his untimely and painful death at the hands of the Tuscarora Indians ended his dream. The attack on Lawson initiated the Tuscarora War (1711–15) in North Carolina. Angered by mistreatment from Indian traders and fearful of increased settlement on Indian land, the Tuscarora fought to oust the Carolinians. With defeat, the tribe eventually left North Carolina, relocated to New York and became the sixth nation of the Iroquois Confederacy.

Ironically, while Lawson was often sympathetic to Tuscarora concerns, his surveying of New Bern and the 1710 settlement there were factors in the outbreak of war. He had met Christoph von Graffenreid in London. Graffenreid (1661–1743), a native of Bern, Switzerland, as part of Georg Ritters Company, planned a new North Carolina settlement and town with Lawson. The proposed settlers were poor Swiss and German farmers whom the English had displaced from the Rhineland (Palatinate) during the War of Spanish Succession. For the settlement, the Georg Ritters Company purchased nineteen thousand acres from the Lords Proprietors. The land lay on the Neuse and Trent Rivers in eastern North Carolina. In 1710, Graffenreid, Lawson and about eight hundred colonists sailed from England. The settlers faced many challenges and setbacks, including a pirate attack, disease and, later, Indian attacks. Graffenreid persisted and named the new settlement New Bern in honor of Bern, Switzerland. Unfortunately for Lawson, Graffenreid and the settlers, the site of New Bern was also the site of a Tuscarora village. Consequently, the influx of these new settlers and the new town aggravated Tuscarora–English relations. The capture of Lawson and Graffenreid was the opening volley in the Tuscarora War. The Tuscarora tortured and killed Lawson but eventually released Graffenreid.

In addition to Lawson's viticultural efforts, the settlement at New Bern has its own significance for the history of wine in North Carolina. The German-speaking settlers at New Bern made wine from the native grapes. The flavor of the wine reminded them of muscadet (a light dry wine) and muscatel (a sweet European dessert wine). Hence, the popular name for the *vitis rotundifolia* grape is "muscadine." The town of New Bern was also the colony's first permanent capital as well as the first capital of the state of North Carolina.

By 1737, there were small vineyards at Bath and New Bern. John Brickell, a visiting physician, found the wine of the latter "exceedingly good." Later, North Carolina colonial governors also sought to promote grape cultivation and wine production. In 1739, Governor Gabriel Johnston, born in Scotland, noted the grapes growing along the Cape

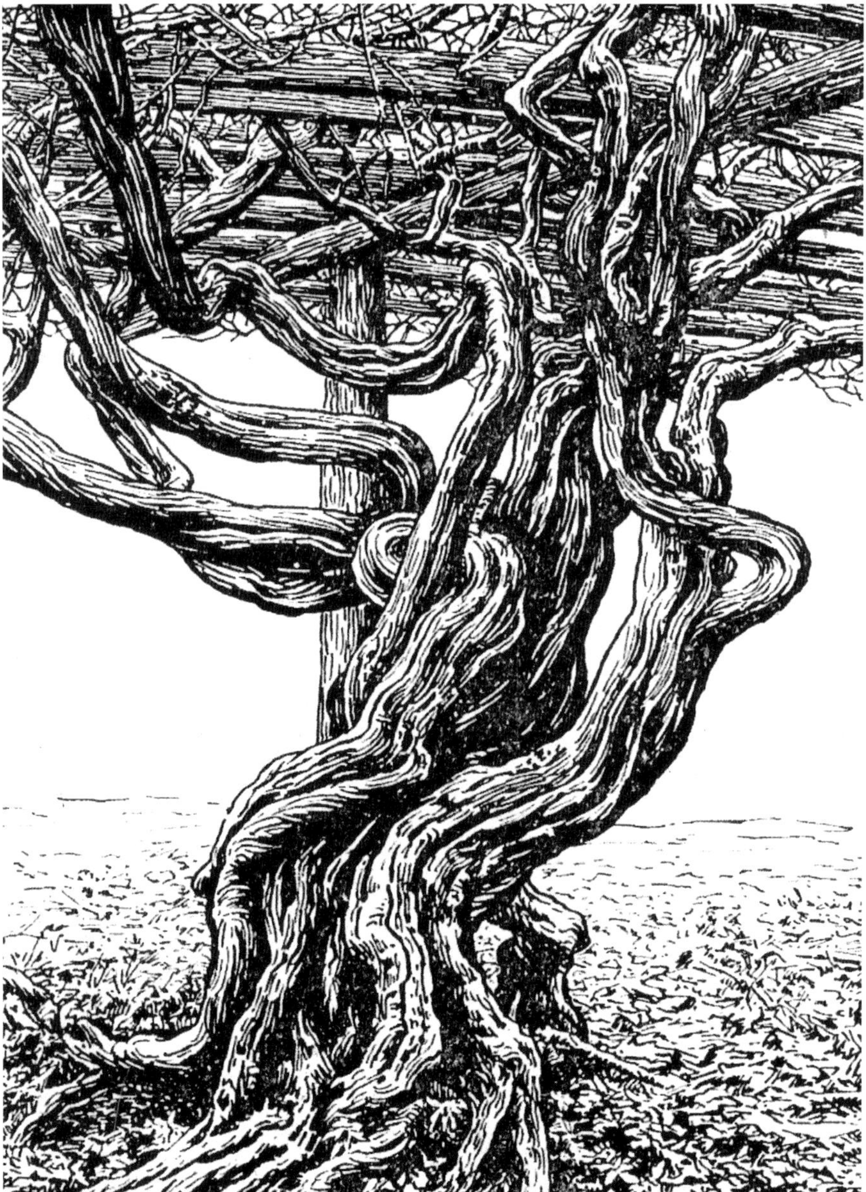

Trunk of an old *rotundifolia* vine (woodcut). The *Vitis rotundifolia* vine lives to a great age. Trunks of the old vines are often solid and tree-like. *George C. Husman,* Muscadine Grapes, *courtesy of the author.*

Fear River and encouraged settlers to grow grapes for wine. Gabriel Johnston (1699–1752) was North Carolina's second royal governor and the state's longest-serving governor; he served from 1734 until 1752. During his tenure, Scots-Irish settlers moved from Pennsylvania, Virginia and Maryland down the Great Wagon Road into the Yadkin Valley. These were critical decades in the peopling of the Carolina Piedmont. In addition, a group of Moravians purchased from Lord Granville the 100,000-acre Wachovia tract in Forsyth County, North Carolina. Their involvement in North Carolina's wine history will be discussed later.

In 1755, Governor Arthur Dobbs (1689–1765), born in County Antrim, Ireland, commented on the "great variety of native Grapes, which yield rich wines" that "only want proper vine Dressers to improve them." Dobbs, a former member of the Irish House of Commons (1729–30), was one of North Carolina's five royal governors, serving from 1754 until 1764. He also had extensive landholdings in the colony, especially in Mecklenburg and Cabarrus Counties. Dobbs encouraged immigration and promoted the economic development of North Carolina—hence his particular interest in grape cultivation as an occupation for the settlers and a product for trade.

During Johnston's tenure as governor, Moravians from Pennsylvania had purchased the Wachovia tract, and in 1753 and 1754, the first settlers arrived. The Moravian Church or *Unitas Fratrum* (Unity of Brethren), an outgrowth of the ministry of John Hus, a martyred Reformer, officially began in the modern Czech Republic in 1457. The early Moravians were frequently persecuted, and several groups eventually settled in the English colonies of North America. Before the American Revolution, the Moravians established three villages and three rural areas in the American colonies. In the 1740s, there were Moravian settlements in Bethlehem and Nazareth, Pennsylvania. The first settlers of the Wachovia tract were from the Bethlehem, Pennsylvania settlement.

The Moravians planned their settlements well. They laid out towns and farmland. Interested in sustainability, they were also pioneers in forest management. In addition, the Moravians who settled in the Winston-Salem area planted vineyards—originally with *vinifera* grapes—for their own use. When the European varieties did not thrive, the Moravians cultivated the native muscadine grapes. In 1769, they reportedly produced nineteen hogsheads of native grape wine.

As an interesting sidelight, Captain John Abraham Collet, a native of Switzerland whose map of North Carolina was published in 1770, also wrote a description of Anson County. In 1769, Collet wrote:

> *The whole back Country…is Covered with pretty good wild grapes, and Vines transplanted from Europe, or other parts of the World, thrive surprisingly, but there are but few; it would be worth while to plant Vine Yards there, for, besides the Immense benefit they would be to the Country, and the Planter, Government has promised a reward of five hundred pounds to the first who shall make a Hogshead of Good Wine.*

If Collet were correct in his assertion, the Moravian vintners may have qualified for the prize. Collet (who died in 1789) came to North Carolina in 1767 as a royal appointee. He served as governor of Fort Johnston near Wilmington and, in 1768, was part of Governor William Tryon's campaign against the Regulators.

During the pre–Revolutionary War years, the colony of North Carolina sought to diversify its economy by exploring several options, including staple crops, naval stores, silk manufacturing and the elusive commercial wine industry. The government seemed equally accepting of native or European grapes if a successful wine was the outcome.

In 1771, Edward Antill (1742–1789), later a lieutenant colonel and British prisoner during the American Revolution, contributed an essay on North American winemaking and grape cultivation to the *Transactions of the American Philosophical Society*—*Essay on the Cultivation of the Vine, and the making and preserving of wine, suited to the different climates in North America.* In the essay, Antill, a native of New Jersey, stressed the importance of vineyards in farming, the use of organic fertilizers and the significance of wine for domestic use and trade. As Antill stated,

> *The planting of Vineyards, the cultivation of Vines, the making of Wine, and Casks to preserve it, must employ and give bread to a great number of people; the freight and profitable remittance, must enrich the merchant; and the being supplied from the colonies with wine, in exchange for her manufactures, must be a considerable saving to Great Britain.*

In addition to his understanding of British mercantilism, he also had high expectations for the wine produced in the United States. As he wrote, "The making of wines the preserving them, will soon arrive at greater perfection, than yet it has been done." He asserted that America had an advantage in that winemakers in the colonies could build on the experience of European vintners and avoid the same mistakes.

In addition to cultivation and winemaking tips, he suggested several varieties of grapes for possible cultivation in the different colonies. For the colonies of Maryland, Virginia and North Carolina, Antill recommended the Chasselas Blanc (a Swiss grape from the Lake Geneva area, perhaps originally from Egypt), known as the Royal Muscadine; the Malvois or Malmsey (Madeira grape); grey and red Frontignac (a grape of many varieties and colors used to make a sweet wine in Frontignan (Languedoc), France); black and white Lisbon; and the Chasselas Noir (an Italian wine grape). The author also suggested revitalizing exhausted land with mud (rich in organic material) from the rivers and creeks.

For colonial and antebellum farmers, a successful vineyard produced grapes for the table, wine for home consumption and grapes and wine for income. Not only could the grapes be sold, but also, when money was in short supply, wine was a medium of exchange.

Vitis Rotundifolia

Domestically, the Indians of the southern colonies enjoyed and perhaps cultivated the muscadine, and from the earliest European settlements, colonists grew them as well. *Vitis rotundifolia*'s natural habitat is the Old South—the states south of the Mason-Dixon line: Virginia, North Carolina, South Carolina, Georgia, Florida, Alabama, Mississippi, Louisiana and Texas. Muscadine vines are at home in the plains and piedmont of the southeastern and southern states. The old Chesapeake colonies of Virginia and Maryland mark the grape's natural northern boundary. Scuppernong grapes and wine have a wonderfully unforgettable flavor. So southerners attempted to capitalize on its commercial possibilities. As simple as the idea seemed, the process was not. Scuppernong grapes,

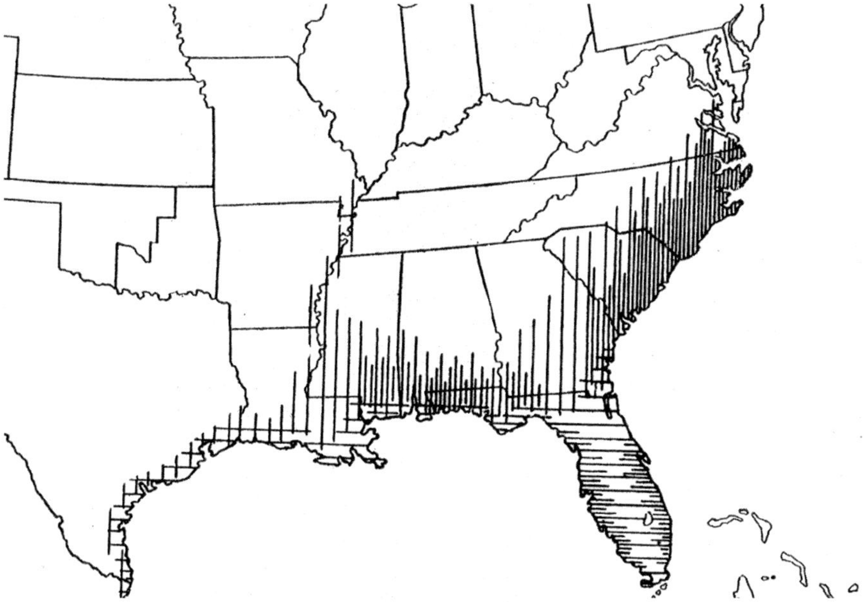

Map of southeastern United States showing natural distribution of muscadine grapes. Muscadine grapes grow throughout the southeastern United States from the James River to Florida. *George C. Husman,* Muscadine Grapes, *courtesy of the author.*

even on the same vine, mature at different times. As a result, harvesting was an ongoing and time-consuming process. Also, in the warm southern states, temperature control was virtually nonexistent. As a result, early winemakers often had difficulty producing scuppernong wines with consistent flavor and alcohol content.

In addition to producing a heavy yield, muscadine grapes are relatively free from disease, insect resistant, not subject to rot or mildew, can survive temperatures of twelve degrees below freezing and are easily propagated. When properly cared for, the vines live to great age, and the muscadine grapes yield a number of noteworthy byproducts—for example, syrups, ketchup, jelly, jams, nutritional supplements and dermatological creams. By 1870, the scuppernong was considered "a delicious table grape" and "the wine grape of America"—"superior to all other native varieties" because the white grape was a "grade richer in sugar, and finer in flavor." Also, in the *Rural Carolinian* of 1870, Louis Froelich of Kenansville, North Carolina, wrote that the scuppernong was "the most profitable grape in the world."

The Scuppernong

By 1913, the scuppernong was the principal grape grown in the Southeast. The name is of Indian origin, but the precise source of the scuppernong variety of the muscadine is not known. Many credit the Scuppernong River, while others cite the Mother Vine on Roanoke Island. Much myth and legend surround the Mother Vine, also known as "Sir Walter Raleigh's Wine." To some, it is the "mother vine" for all other scuppernong vines—the oldest scuppernong vine in North Carolina.

In 1907, Professor C.C. Newman of Clemson University considered the scuppernong "probably the oldest American grape in cultivation." Some credit its discovery to the English settlers of the Lost Colony on Roanoke Island; others to a site near the Scuppernong River. J.D.

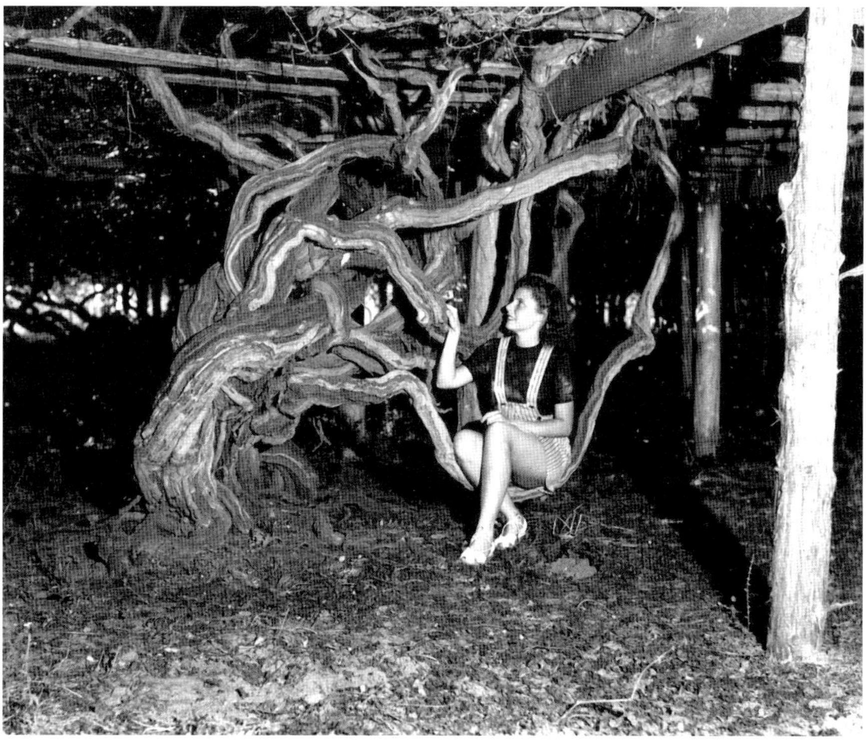

Mother Vineyard, historic scuppernong vine, Manteo, North Carolina. Some believe that this vine is the oldest scuppernong vine in the word. *Courtesy of the North Carolina State Archives.*

Cameron writing in the North Carolina *Handbook* in 1893, stated that the scuppernong was known before 1774, when the Reverend Mister Charles Pettigrew found it growing along the banks of the Scuppernong River in Tyrell County. Cameron also included another discovery story for the scuppernong that credited Isaac Alexander with finding the grape in Tyrell County in 1755.

Charles Gohdes, in his history of the scuppernong, recounted the statement of Dr. Calvin Jones in 1817 that he and Thomas Henderson Jr. had designated the grape "Scuppernong" in honor of James Blount of Scuppernong. Jones and Henderson edited the Raleigh newspaper, the *Star*, and Blount had frequently written to the paper about the grape. Jones was a veteran of the War of 1812 and a trustee of the University of North Carolina–Chapel Hill. In 1810, Blount reported that residents of Washington County had produced 1,368 gallons of wine, primarily from the scuppernong grape.

The Mother Vine

According to George Husman, writing in 1913, the old scuppernong vine growing on Roanoke Island, known as the Mother Vine, was "very old" but not the original vine found by the colonists. However, some believe it could be a descendant of the Roanoke Island vine. F.C. Reimer, another horticulturalist, echoed those sentiments. From Reimer's perspective, the oldest scuppernong vines in North Carolina grew in Tyrell County. His finding that the five vines on Roanoke grew in straight lines equidistant from one another suggested that they were the remnants of an early—perhaps, in Charles Gohdes's opinion—1850s vineyard. This 1850s date could link the vine with the first commercial vineyard known on Roanoke Island.

Regardless of the possibilities, the Mother Vine is an old *rotundifolia* vine. In 1870, D.H. Jacques writing in the *American Naturalist*, reported that "a vine, believed to be the original Scuppernong, and 300 years old, is now growing on Roanoke Island, N.C., which is said to have yielded last year forty-six barrels of wine." David W. Barefoot described it as a

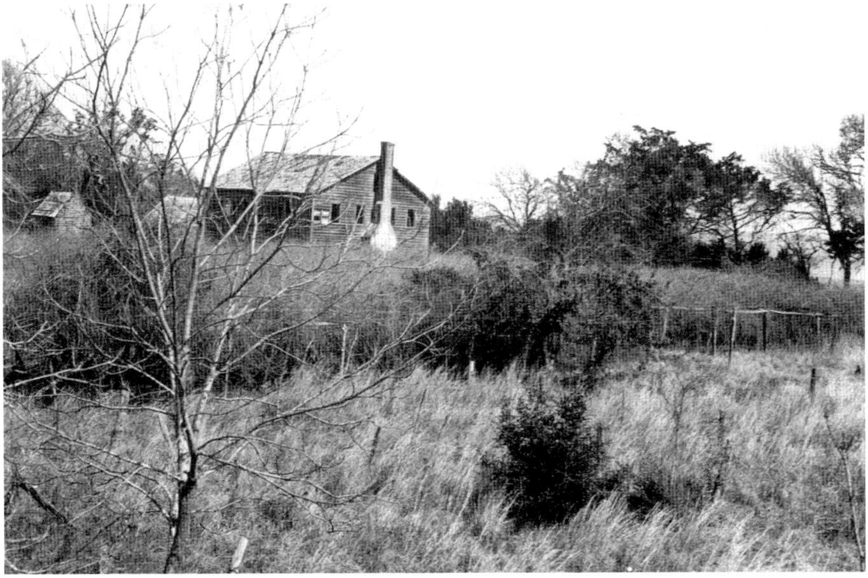

Baum House, Mother Vineyard with grape arbor, Manteo, North Carolina. Peter Baum, the progenitor of the Roanoke Island family, reportedly owned this property by 1729. *Courtesy of the North Carolina State Archives.*

"gnarled and twisted vine." Questions persist about who planted the vine and when. According to the *North Carolina Reader* of 1858:

> *The first vine of this name* [scuppernong] *was found on the banks of the Scuppernong River…by some of the party composing of the first Anglo-Saxon Settlement on Roanoke Island…one small view was transplanted very soon in Roanoke Island, where only a few years ago I was told by old Abraham Baum, then 84 years old, that when he was a boy the vine was the largest on the island.*

Solomon Baum (1813–1898) was the son of Maurice Baum (1772–1839) and the grandson of Abraham Baum (1742–1833). Peter Baum, the father of Abraham, supposedly acquired the site of the Mother Vine before 1729. In 1850, Solomon Baum was a farmer in the Roanoke Island District of Currituck County with real estate worth $627. In 1860, he still lived on Roanoke Island. However, in 1870, he was listed in Nags Head, Dare County. On other occasions, Solomon Baum described the

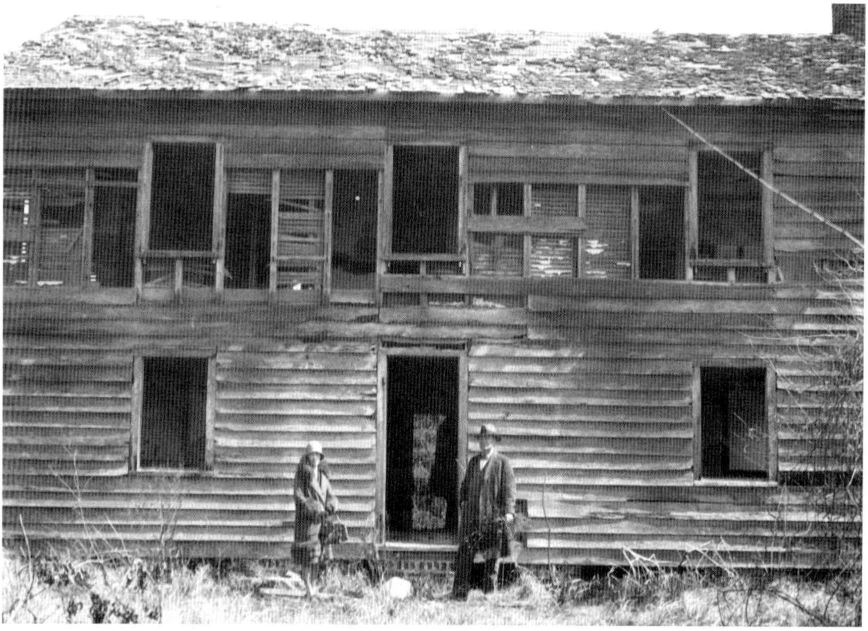

Maurice Baum House, Mother Vineyard, Roanoke Island. Maurice Baum (1772–1839) was the father of Abraham Baum, who remembered the Mother Vine "as the largest on the island." *Courtesy of the North Carolina State Archives.*

vine's trunks "as big and old" and noted that the "vine had not changed during his lifetime."

Some disregard the possibilities of a mid-nineteenth-century vineyard or the hard-pressed Roanoke settlers planting grapes. Instead, according to Ray Py, the Native Americans of Roanoke Island may have planted the vines.

The scuppernong is but one variety of muscadine (*Vitis rotundifolia*). Muscadine grapes are variable and prolific. These grapes bear fruit in clusters of three to eight berries, each of which may be from one-half to one inch in diameter. In 1920, there were thirty-five to forty named varieties of muscadine grapes. According to Husman, "The cultivated varieties are accidentally discovered natural seedlings of wild vines having superior qualities." In 1913, the leading varieties of muscadine were the scuppernong, Flowers, James, Thomas, Mish, Memory and Eden. The scuppernong is a light-colored variety of *rotundifolia* valued for eating and winemaking. In addition to the scuppernong, most of these grapes are native to North Carolina. For example, "Popping Billy" Flowers discovered the Flowers grape, a purplish black variety of *rotundifolia*, near Lumberton, Robeson County, North Carolina. By 1914, the Flowers grape, a late-blooming grape, was more often used in winemaking than for home consumption. In 1866 or 1867, B.M.W. James of Grindool found the James variety near Grindool Creek in Pitt County, North Carolina. The James variety has large, round grapes and was a popularly cultivated dark *rotundifolia*. It is the oldest named dark variety of *rotundifolia*. Between 1840 and 1850, Albert Mish discovered the Mish grape growing near the Pamlico River. In 1850, the sixty-seven-year-old Mish, a native of Germany, lived in Beaver Dam Township, Beaufort County. These are a few of the many varieties of muscadine discovered and cultivated in North Carolina before 1900. Several of them formed the backbone of North Carolina's early twentieth-century grape industry.

Chapter 2

Medoc and Commercial Success

The best grape in the world.
—Sidney Weller

The first vineyards in North Carolina grew grapes and produced wine for local trade and home consumption. Among those early producers was John Grady. At the time of the American Revolution, Grady of Albertson in northern Duplin County crushed grapes and made muscadine wine. Grady, who died during the Battle of Moore's Creek Bridge in 1775, was the first North Carolina casualty of the Revolutionary War. The Battle of Moore's Creek Bridge, fought on February 27, 1776, was a significant milestone in the history of the American Revolution in the South. According to North Carolina historian William S. Powell, many consider the battle "the Lexington and Concord" of the southern colonies. It was an important early Patriot victory that gave the independence movement time to develop. A monument at the battle site honors Grady's sacrifice.

These early growers prized the white muscadines—especially the scuppernongs, the "great summer or bull-grape of Carolina"—for their sweet wine. Early wine producers included Ebenezer Pettigrew of Edenton, Chowan County, and George E. Spruill of Plymouth, Washington County, who owned a famously large scuppernong vine. Pettigrew (1783–1848) was a well-to-do planter in the Lake Scuppernong

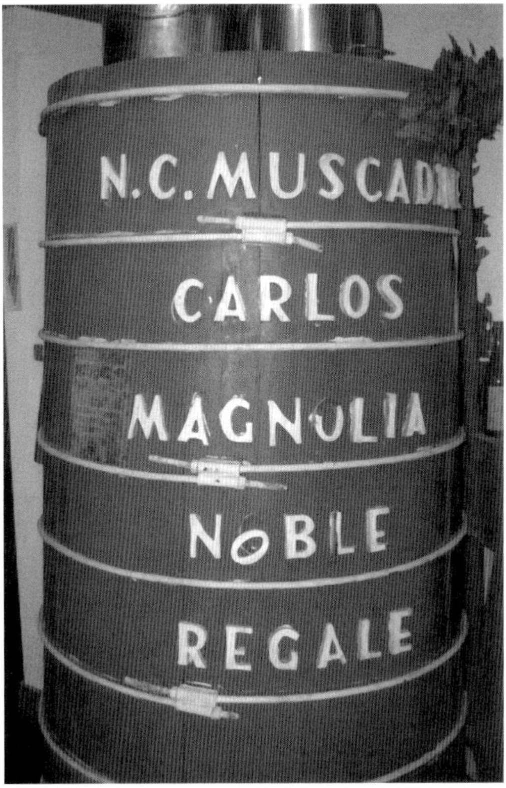

Early North Carolina muscadine grape crusher, circa 1776. Early North Carolina settlers generally produced wine for local or home consumption. John Grady of Albertson in Duplin County owned this piece of equipment. Grady died during the Battle of Moore's Creek Bridge in 1775, the first North Carolina casualty of the Revolutionary War. *Photograph by Terry L. Helsley, courtesy of David G. Fussell Sr.*

(Phelps) area. He attended the University of North Carolina–Chapel Hill and served in the North Carolina State Senate from 1809 to 1810 and in the Twenty-fourth United States Congress (1835–37). In 1835, Pettigrew wrote a friend, "My business is going well…I have been for the past week making wine."

Even that noted wine connoisseur Thomas Jefferson drank and praised North Carolina scuppernong wine. In his correspondence with John Bartram (1699–1777), the noted naturalist and explorer, known as the "father of American botany," Jefferson advocated the increased consumption of wine as a healthier alternative to whiskey. However, he was selective in his suppliers, as he wanted to avoid the wines heavily fortified with brandy. Consequently, in 1822, Francis Wayles Eppes, Jefferson's grandson, wrote Jefferson recommending Pettigrew and Spruill as the producers of the best scuppernong wine in North Carolina. Eppes also commented that "the former [Pettigrew] will not always sell, being very wealthy, the latter is not in as good circumstances, and owns the famous vine covering an acre of ground."

According to winemaker David Fussell, there were two reasons to add brandy to wine. First, the alcohol killed germs and prevented spoilage. Second, adding applejack brandy to grape juice produced wine without fermentation or aging—a convenience for wine producers who wanted to get a product to market quickly. Consequently, scuppernong wine, according to Thomas Pinney, was for many years simply "fortified grape juice."

As an alternative wine-producing method, other producers let the crushed grape must (juice, pressed fruit and stems) ferment and then mixed it with brandy. Some winemakers mixed the juice with honey, while others preferred the wine without additives.

Jefferson, the third president of the United States, developed a taste for French wine while he served as America's minister to France. During his first term as president, Jefferson spent a princely sum—$7,500—on wine. He was not only famous for his wine cellar (in 1985, a bottle of wine reputedly from Jefferson's Paris cellar brought $125,000 at auction) but also as a viticulturist. At Monticello, his Virginia estate, he planted several vineyards primarily of *vinifera* grapes. He attempted to avoid the *vinifera* pest problem by grafting *vinifera* vines onto native roots. In many ways, Jefferson was a man ahead of his time. Only hybridization, grafting and modern pesticides finally made the commercial cultivation of *vinifera* in the United States possible. In addition, Jefferson backed several other unsuccessful efforts to produce wine from native and European grapes in the United States. Despite repeated failures and disappointments, he never lost his vision of America as a grape-growing, wine-producing country.

In 1825, John S. Skinner, editor of the *American Farmer*, printed a letter from Plymouth in Washington County, North Carolina, about scuppernong wine. The unsigned letter raised the soon-to-be familiar argument that wine drinking was more "temperate" than drinking whiskey and other liquors. The writers stated that at that time the winemakers produced three to four hundred casks of wine annually. The contributors also sent Skinner two dozen bottles of scuppernong wine—one dozen for his personal consumption and one dozen to exhibit at the Maryland agricultural fair. This letter also described scuppernong wine production as follows: "This wine is made without any fermentation, but simply by pressing the grapes, and by mixing 3 gallons of the pure juice with 1 gallon of apple brandy." According to the author, in 1825 scuppernong growers could earn a profit of $400 per acre.

In 1870, D.H. Jacques—an American author who in 1871 edited the *Rural Carolinian*, an illustrated agricultural magazine published in Charleston, South Carolina—described the scuppernong as "often tender and deliciously sweet." The *Rural Carolinian* was a Grange (the National Grange of the Order of the Patrons of Husbandry) publication printed by Walker, Evans & Cogswell and D. Wyatt Aiken. Jacques also noted that "in the best specimens the pulp finally dissolves, and skins become literally bags of wine."

In addition to the muscadine, North Carolina has a number of other distinctive native grapes. These grapes include the Lincoln, a variety of summer grape that originated on the Catawba River in Lincoln County, and the fox grape (*Vitis labrusca*). Settlers found fox grapes in the middle and western sections of the state. Perhaps the earliest identified variety of the fox grape was the Isabella grape. As early as 1805, North Carolinians were cultivating the Isabella near Wilmington. In western North Carolina, the Catawba grape originated on Cane Creek in Buncombe County. During a meeting of the boundary commissioners from North and South Carolina at William Murray's Inn in Fletcher, North Carolina, one of the North Carolina commissioners took clippings of a wild grape found there. About 1801, Captain Clayton popularized the grape. So, plantings of the grape traveled from Buncombe County to Catawba (at that time Lincoln) County. From there, growers took the grape to Cincinnati, Ohio, and finally to Lake Erie, New York.

The Catawba grape is also historically significant for the wine industry in America. In 1828, Nicholas Longworth, a wealthy banker, merchant and attorney in Cincinnati, Ohio, built a successful wine-producing industry with the Catawba grape. Longworth (1783–1863) was one of America's winemaking pioneers. Indeed, to some, he was the father of American grape culture. Using the Catawba, a North Carolina grape, he proved that Americans could successfully make and market American-made wines from American native grapes.

Longworth's success encouraged other would-be wine producers to reexamine the possibilities of American native grapes. Longworth's grandson, also named Nicholas, was Speaker of the United States House of Representatives and married Alice Roosevelt, oldest daughter of President Theodore Roosevelt.

Major John Adlum in the District of Columbia also contributed to the Catawba's reputation and the early American grape industry. Adlum, a veteran of the American Revolution, found the grape in Maryland, cultivated it and sent clippings to Longworth. According to wine historian Leon Adams, Adlum gave the grape its name and commented to Longworth that by focusing attention on this grape he had rendered the United States a "greater service" than if he had "paid off the national debt."

From the Catawba, grape growers developed many other varieties. Among the popular Catawba cultivars were Concord (a major source of produced grape juice), Hartford, Prolific, Niagara and Martha grapes. Catawba grapes were great table, wine and juice grapes and were frequently used for jams and jellies.

Despite the potential of the native grapes—*Vitis labrusca* and *Vitis rotundifolia*—widespread grape cultivation and winemaking developed slowly in the South. In 1811, an unspecified North Carolina county produced over 1,300 gallons of wine, and by 1819, a number of farmers produced wine for home consumption and sale along the Cape Fear River. An 1817 visitor to Roanoke Island named Samuel Huntington Perkins noted that the island inhabitants made good wine but exported little, retaining it instead for local consumption.

Interestingly, it was a transplanted northerner who first brought commercial success to the North Carolina wine industry. As a result of that breakthrough, other farmers, particularly in eastern North Carolina, turned to grape growing. Therefore, by 1840, there were so many scuppernong wine producers in North Carolina that the state led the rest of the United States in wine production. Yet winemaking was primarily a coastal phenomenon.

Medoc

Brinkleyville was the first known center of commercial wine production in North Carolina. The town of Brinkleyville, located in Halifax County, was named for Joseph Brinkley, a colonial settler. Brinkleyville had a post office by 1817, and nearby Ringwood had a school in 1800. In 1829, Sidney Weller (1791–1854) moved to the Halifax area. Weller was a complicated man with many interests and personae, and in addition, he was a talented self-promoter. At times, he advertised as a teacher, a headmaster, a preacher or an agricultural innovator. Regardless, he was an agricultural pioneer who advocated and practiced scientific farming and crop rotation. He was also a critical factor in the commercial success of what had been a mainly domestic manufacture. He similarly promoted the planting of Chinese mulberry trees and tried to jump-start a silk industry in North Carolina. The mulberry trees thrived, but the silk venture, though widely publicized, did not succeed.

Born in New York State, Weller, an experienced educator and graduate of Union College, relocated to Virginia and, shortly thereafter, settled with his wife in Halifax County, North Carolina. There, after filling a post at Quankey Academy, Weller opened his own school, Brinkleyville Academy. In addition to his educational pursuits, in 1829 Weller purchased three hundred acres of worn-out land and began farming using scientific techniques. Weller was not interested in becoming a planter but, rather, in general farming. Weller studied the cultivation of artichokes, okra, rutabagas, corn and other vegetables. In addition to such crops, Weller was cultivating grapes using the methods of Nicholas Herbemont of Columbia, South Carolina. The French-born Herbemont (1771–1839) was a pioneering viticulturalist who began producing wine in 1823. By 1836, he had perfected the art of winemaking, and his writings also influenced other winemakers. Herbemont's farm, Palmyra, was outside Columbia on the road to Camden, South Carolina.

In Halifax County, Weller not only grew grapes in his nursery but also studied winemaking. In 1835, Weller had his first commercial wine production. In the beginning, he shipped his wines from Weller's Vineyard, but later he styled his vineyard as Medoc Vineyard.

Although not the first vineyard in Halifax County, Weller's vineyard was an important one. He promoted grape cultivation and scientific agriculture, and his vineyard would have long-term significance for the economy of antebellum and modern North Carolina. By 1840, Weller's vineyard, though small by today's standards (six acres), was the largest in North Carolina. By 1848, Weller noted that his vineyard had grown to eight acres. He shared his ideas on grape cultivation and promoted the enterprise. In 1848, he told *DeBow's Commercial Review* readers that "land that will bring good corn is rich enough for vines." Other advice for potential vineyard planters included having a "southern exposure," "light soil" and a "high and dry situation."

Today, Medoc Mountain State Park includes part of Weller's property. The park was established in the 1970s. Medoc is not a mountain but the remnant or core of a mountain range formed during the Paleozoic Age. Weller chose the name because Medoc was a famous grape-growing region in Bordeaux, France, recognized for its red wines. A North Carolina Historical Marker, erected in 1982 at the state park, honors Sidney Weller and his contributions to grape culture in North Carolina.

Perhaps influenced by Captain William Burlingham, a grape grower and winemaker near Louisburg, Weller cultivated scuppernong grapes. According to Weller, Burlingham had twelve vines of scuppernong grapes that produced five hundred gallons of wine. While he bragged of cultivating hundreds of varieties of grapes, Weller never lost his major interest in cultivating native grapes. In addition, he developed new varieties of muscadines, such as the Halifax, or "Weller's Halifax." Weller did not adopt European methods of grape cultivation but, instead, developed the American method. He grew his grapes on scaffolding and did not prune them. As he wrote, "The very luxuriant, far-extended growth of grape vines in our forests and hedge rows is the true model of nature, pointing out the unerring plan for an American vine grower to pursue." A scuppernong vine can live forty or more years and grow to a great circumference.

By 1840, Weller's Medoc Vineyard was flourishing. In September 1846, Weller made thirty-three gallons of wine from five bushels of white scuppernong grapes, two bushels of Bland's Virginia grapes (a red muscadine) and two and a half bushels of wild bunch grapes. He sold the

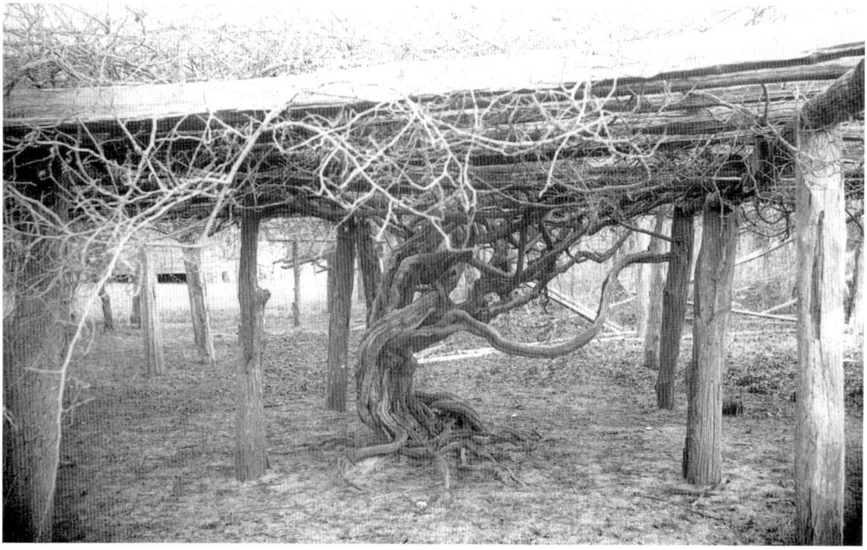

Old scuppernong grapevine, Roanoke, circa 1900. Scuppernong vines require scaffolding to support them. Old vines reportedly can shade half an acre. *Courtesy of the North Carolina State Archives.*

resulting wine under the name of "Weller's Scuppernong Champagne." Not only did Weller produce and sell wine, but he also sold grapes and grape plants. In addition, he exhibited his grapes—some with a diameter of four inches—at the first North Carolina State Fair in 1853. He contended that three-and-a-half-inch grapes were fairly common. In addition, he contributed articles on grapes and their cultivation to agricultural journals.

In an article for *DeBow's Commercial Review* abstracted in the *American Agriculturalist* of 1848, Weller stated that one year he made forty barrels of wine that sold for between one and four dollars per gallon. For example, scuppernong Madeira sold for three dollars per gallon and scuppernong Hock (a red wine) for four dollars per gallon. To produce his lower-end wines, Weller added brandy, while for his higher-end wines he added refined sugar. Weller's vineyard included his favorite scuppernong grapes, as well as Weller's Halifax (a cultivar developed by Weller), Norton's Virginia Seedling, Le Noir, Hunterville, Franklin, Bland's Virginia and other native cultivars. Weller reported that a fourteen-year-old scuppernong covered fifty feet, and on average, in the

Halifax area, scuppernong vines produced two thousand gallons of wine per acre. Reputedly, one of Weller's scuppernong vines alone produced fifty gallons of wine annually. Weller also produced and marketed White Scuppernong Hock, Weller's Halifax and a scuppernong champagne.

According to H.G. Jones, Weller had been involved in the organization of the State Agricultural Society of North Carolina in 1852 and, as a result, was involved in the holding of North Carolina's first state fair. Shortly before his death, Weller asserted in the *Farmer's Journal* that for

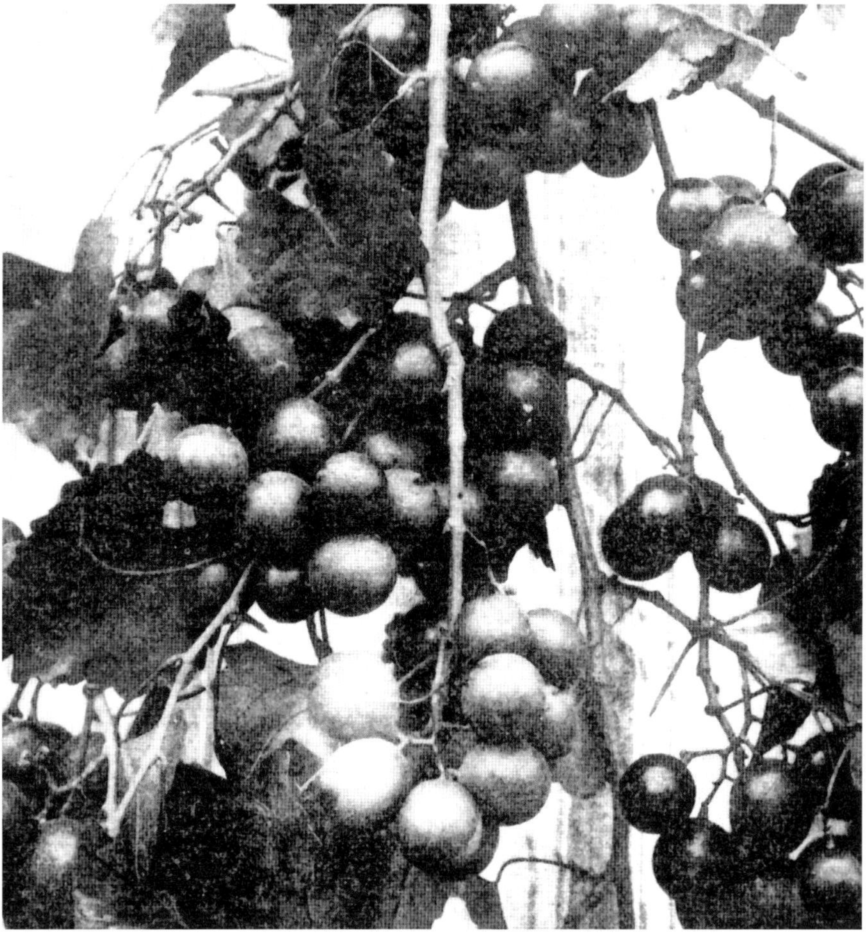

Fruit and branches of *Vitis rotundifolia*. Muscadine grapes do not grow in bunches. Rather, they bear fruit in clusters of from three to eight berries. *George C. Husman,* Muscadine Grapes, *courtesy of the author.*

North Carolina's climate, the scuppernong was the "best grape in the world." His wine was sold nationally, and like later vintners, he made his vineyard a destination. Weller opened the vineyard to the public for a fee and encouraged picnicking. In addition, he stressed the healthfulness of his scuppernong wine. The secessionist Edmund Ruffin called one of Weller's vintages "the most delicious drink we have ever enjoyed." In 1850, the fifty-nine-year-old Weller lived with his wife, Eliza, and their eight children in Halifax County. At his death in 1854, Weller owned the largest vineyard in North Carolina.

According to the 1860 Agricultural Census of Halifax County, in addition to Weller's family, several other individuals—including Phoebe Pullen, J.J. Hervey and Thomas Wilcox—also produced wine in Halifax County. Pullen, Hervey and Wilcox all appear on the 1860 population schedule as farmers. Pullen and Wilcox, like Weller's family, listed Brinkleyville as their post office. J.H. Hervey, age seventy-seven, was served by the Ringwood Post Office. As North Carolina was the leading wine-producing state in the United States in 1840, many others had to have been involved. For example, Charles L. Pettigrew's Pettigrew Vineyard on Lake Scuppernong (Phelps) in Tyrell County was in operation before 1850. In 1850, Pettigrew was a farmer with real estate valued at $22,000. Similarly, the fifty-acre Cannon Farm near Chapel Hill, North Carolina, began wine production as early as 1850 or 1860, and F.S. Proctor planted a vineyard on Roanoke prior to 1859. The 1860 census listed F.S. Proctor as a forty-eight-year-old farmer with real estate worth $3,600 and a personal estate of $26,750. Proctor lived in Pasquotank County, with his post office in Elizabeth City.

The Weller saga ended in 1867 when Elizabeth, his widow, and their older son, John, sold the vineyard and winery to the Garrett brothers, Dr. Joseph J. Garrett, Charles W. Garrett and Dr. Francis M. Garrett. However, much of Weller's pioneer work survived. The scuppernong is still cultivated in eastern North Carolina. Modern wineries seek to capitalize on the natural beauty of their vineyards and their admirable wine with tours, picnic areas, outdoor concerts and tasting rooms.

The Garretts, though, would take the North Carolina wine business to a new level. Charles William Garrett managed the vineyard called Ringwood, and under his leadership, Ringwood Vineyards expanded and annual production increased. The Garretts used most of the grapes produced for wine manufacture. In some seasons, the vineyard yielded 175,000 gallons of wine. Under the Garretts, the vineyard produced table wines, sparkling wine and brandy. The continuation of the Garrett story will take North Carolina wines from regional to national and international prominence.

Early *Vinifera* Production

Shortly before 1850, a new wine enterprise challenged the scuppernong's supremacy. Dr. Joseph Togno, who reportedly had cultivated *vinifera* grapes in Fauquier County, Virginia, established a vineyard and school on Diccoteaux, his plantation near Wilmington, North Carolina. While a medical student in 1827, Togno had translated a French medical text. In 1828, his fiancée, Frances Rundle, died in Philadelphia. Later, in 1840, he was a tutor or perhaps professor of modern languages at the University of Virginia. On the 1850 census, the French-born Togno, age fifty-three, is listed as a vineyard tender in the county of New Hanover, North Carolina. In addition, Togno wrote on grape cultivation and the history of native wines and contributed to Charleston, South Carolina newspapers and the United States Patent Office reports. For the latter, he wrote a long letter on the history and origins of the Isabella grape. *Putnam's Monthly* in 1854 included the following statement: "Mr. Togno is sanguine as to the introduction of the foreign grape in this country for open vineyard culture." However, neither the "Vine Dresser Model School" nor the *vinifera* grape plantings succeeded, and in 1853, Togno closed his school and left the area. While not despairing of European grapes, he did appreciate the potential of the scuppernong as a champagne grape. On February 5, 1859, according to his obituary, Dr. Joseph Togno "fell dead in the porch of his residence at Abbeville, South Carolina." The obituary also listed his birthplace as Corsica.

As a result of Togno's experience and other failed efforts to successfully grow European grapes, many horticulturalists and viticulturists in the antebellum United States considered *vinifera* grapes poor risks for commercial wine production. European grapes in America were prey to disease and insect infestation. Also, North Carolina's climate made cultivation difficult. The temperate climate meant that the nights were too warm in the summer, rain fell at the wrong time and late frosts often damaged early spring growth.

In October 1860, however, Dr. Francis Joseph Kron (1798—1883), a native of Prussia, wrote Thomas Ruffin, a noted North Carolina jurist, that he was sending six bottles of wine as an exhibit for the North Carolina State Fair. Kron, a physician and horticulturalist, lived in Albemarle Township, Stanly County, North Carolina, and had, at one time, been an instructor at the University of North Carolina–Chapel Hill. He immigrated to America in 1823 and was the first doctor to practice in the southern North Carolina Piedmont. In 1850, his household included his wife, Mary, and their daughters, Adelaide and Elizabeth. He owned real estate valued at $7,000. In his letter to Ruffin, Kron asserted that the bottles contained wine from the "Lenoir, alias *Warren*, alias *Herbemont*, alias *Lincoln*" grape. Still, Kron considered the wine of French origin and wrote Ruffin that the grape used produced a "superior table wine" and that when aged, the wine compared with Madeira. In a footnote, Kron added that he had planted 104 new French varieties in 1859 and 20 of them had produced fruit. Today, the site of Kron's homestead is part of Morrow Mountain State Park in Albemarle. Visitors can tour reconstructed models of Kron's house, office, infirmary and greenhouse at the foot of Fall Mountain.

The number of vineyards in North Carolina and the other southern states expanded. As a group, the growers were interested in new grape breeds and enhanced cultivation techniques. They organized and shared information in various agricultural journals. In 1860, the Southern Vine-Growers' Convention met at Aiken, South Carolina. Attendees developed a standard system of wine labeling and opposed adulteration of southern wines. Seeing the commonalities of grape growers and winemakers across the United States, the group planned to invite growers across the country to the next meeting. The Civil War, however, intervened, and this attempt at national unity and cooperation also ended.

The antebellum years, then, were successful times for North Carolina grape growers and wineries. North Carolina was finally fulfilling the promise dreamed by Lawson, the Lords Proprietors, governors and settlers. On the brink of war, most southern plantations had vineyards, and wine was a popular beverage. Before the Civil War, North Carolina had at least twenty-five wineries located in eastern North Carolina. These eastern wineries generally cultivated native grapes, especially *Vitis rotundifolia*. Sites of antebellum wineries included Cumberland County (at Tokay), Halifax County (at Medoc), Richmond County (at Cognac), Moore County (at Niagara and Vina Vista) and Catawba County (at Catawba). About 1859, F.S. Proctor of Edenton established the earliest commercial vineyard and winery on Roanoke Island. Proctor cultivated scuppernong, concord and Isabella grapes. *The North Carolina Planter* of 1859 reported that Proctor had three thousand vines. The 1860 census listed Proctor as a forty-six-year-old farmer living alone in Pasquotank County, with his post office in Elizabeth City. The Virginia-born Proctor owned real estate worth $3,600 and personal property, including fifteen slaves, valued at $26,750.

These years of promise were also times of competing grape production and competing dreams. Other generations also pursued Lawson's quest to successfully grow European grapes. The *vinifera* dream, although still unrealized, persisted. This *leitmotif* of wanting to grow the difficult grape when the muscadine easily flourished is a persistent thread in the North Carolina wine story.

Chapter 3

Virginia Dare and Friends

For a drink superlatively fine, try Garrett & Co.'s Escapernong Wine.
—Garrett & Co., Weldon, North Carolina

Secession ended this first flowering of the North Carolina wine industry. The Civil War years were difficult for North Carolina grape growers and winemakers. The war impacted manpower and financing and disrupted markets. Some wine producers left the state; others ceased operations or struggled to survive. Also, the Confederacy and state governments discouraged the importation and manufacture of spirits in order to conserve valuable cargo space and grain.

Yet the immediate postwar years saw new possibilities and new investments. The harbinger of this paradigm shift was the Garrett family. In 1867, three Garrett brothers bought Sidney Weller's Medoc Vineyard. The post–Civil War rebirth of the North Carolina wine industry focused mainly on the native grape. Writing in 1872, C.V. Riley, an entomologist, stated, "Nearly all the varieties of the European grape-vine (*Vitis vinifera*) have, in the end, proved valueless when introduced and cultivated in the eastern half of the United States."

J.S. Cannon, reporting on North Carolina scuppernong grape production at the first annual Fair of the Cape Fear Agricultural Association, circa 1869, had words of admonition for North Carolina winemakers. While extolling the value and ease of grape cultivation and how it compared favorably with cotton production, Cannon also criticized the failures of North Carolina winemakers. Specifically, he contended that failures were not attributable to the cost of growing grapes but rather to two factors: "first, adding sugar to the juice of the grape; second, asking too high a price for the wine." According to Cannon, a consumer had to pay at least two dollars a bottle for a North Carolina wine in a hotel and twenty-five cents per glass in a saloon. Such pricing, in Cannon's opinion, reduced the volume of wine sales. He suggested that North Carolina grape growers and wine producers lower their prices.

Yet, in time, North Carolina wineries rebounded. According to the North Carolina Department of Commerce, post–Civil War North Carolina had wineries at Castle Hayne (New Hanover County), Conover (Catawba County), Eagle Springs (Moore County), Edenton (Chowan County), Gibson (Scotland County), Icard (Burke County), Littleton (Halifax County), Louisburg (Franklin County), Manteo (Dare County), Murphy (Cherokee County), Peachland (Anson County), Pettigrew State Park (Washington County), Holly Ridge (Onslow County), Samarcand (Moore County), Tryon (Polk County) and Warrenton (Warren County).

Garretts and Medoc

After the end of the Civil War, North Carolinians slowly returned to winemaking. As a result, the 1870s were a time of growth for the North Carolina wine industry and of scuppernong wine. In 1876, the *Rural Carolinian*, an agricultural journal at that time published in Cokesbury, South Carolina, carried the following recipe for scuppernong wine:

> *Gather the grapes when fully ripe and crush them in a tub or trough with a common maul, after which press the juice out as close as possible—i.e. as long as it will run; put two pounds of the best white sugar to each gallon of juice; fill a cask brimming full, and leave the bung open. When fermentation ceases, stop the bung in barrel tight; it will take about twenty days to ferment. Bore a spigot hole in barrel with gimlet,*

and put spigot in it, and try it frequently by taking spigot out to see if fermentation is entirely over.

In February or March rack off into clean casks, letting the bungs remain loose for two weeks, so that the gas from the slight second fermentation may pass off; then bung tight and keep till Fall, when it will be fit to drink.

In addition to the Garrett enterprises, there was a scuppernong winery in Columbus County; the Bear Winery in Wilmington, New Hanover County; Colonel Wharton's Tokay Vineyard near Fayetteville, Cumberland County; and others in Raleigh, Wake County; Ridgeway, Warren County; Tryon, Polk County; and at other locations around the state. The number of vineyards and wineries continued to increase as the century drew to a close.

The Garrett winery, however, would have the greatest impact on the late nineteenth- and early twentieth-century wine industry in North Carolina. Three Garrett brothers purchased Medoc Vineyard from Sidney Weller's heirs in 1867. Yet it was Charles W. Garrett who was most concerned with the daily operation of the vineyard. In the early years, he operated as C.W. Garrett & Company. Later, Garrett renamed the winery as the Ringwood Wine Company in Enfield, North Carolina. Although still a proponent of the scuppernong, Charles Garrett diversified the grape varieties and produced more varieties of wine and brandy. He also greatly enlarged the physical plant and the winery's output.

In 1877, Paul Garrett, the teenaged son of Dr. Francis Marion Garrett, joined the firm. Charles Garrett, born in Edgecombe County, was a young man who wanted to learn the wine family business. Perhaps surprisingly, he was a natural promoter and marketer. He worked for his uncle and studied all aspects of the vineyard cultivation and winemaking. Charles Garrett marketed his wine not through the print media, but with traveling salesmen. His first salesman, Henry Spooner Harrison (1852–1937), in a sense, changed the face of the American wine industry. Harrison was a successful salesman and the successful suitor of Charles Garrett's daughter Mary. His real significance, perhaps, is that his involvement in the family enterprise eventually drove Paul Garrett to launch his own business.

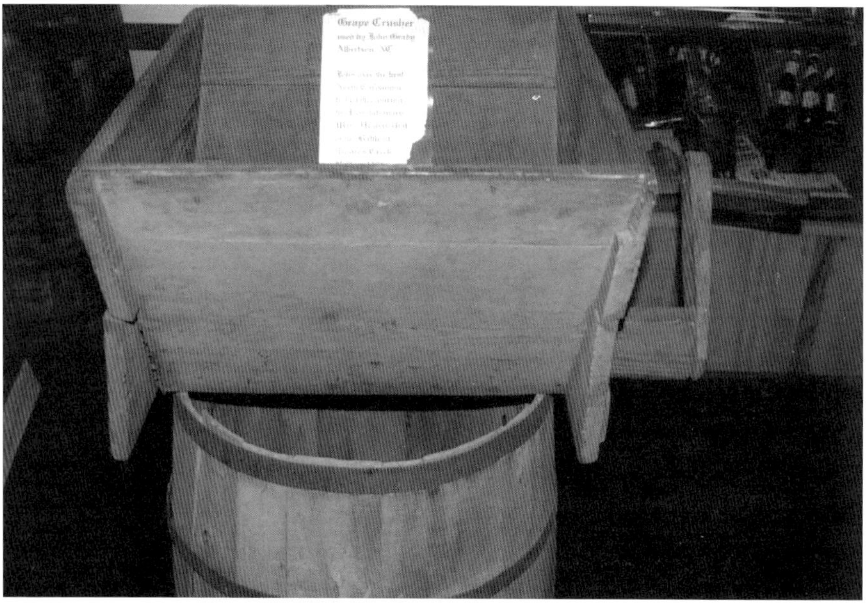

Garrett grape crusher. Paul Garrett expanded his business from distributing wine to producing wine. He owned vineyards in North Carolina and also purchased grapes grown in other eastern North Carolina vineyards. *Photograph by Terry L. Helsley, courtesy of David G. Fussell Sr.*

In 1886, Medoc acquired a post office, and Charles Garrett renamed the winery Medoc Vineyard, with C.W. Garrett & Company as proprietors. In November of that year, Charles W. Garrett died, perhaps with the expectation that Paul Garrett, his nephew and apprentice, would manage the business. That is not how events unfolded. Harrison determined to manage the plant and the vineyard, and Paul Garrett became the traveling salesman.

Paul Garrett had a flair for sales, but a contract dispute motivated him to strike out on his own. In 1900, Garrett negotiated a favorable contract to handle all of Medoc Vineyard's wine at a set price. He negotiated too well, and his initial success stunned Harrison, who refused to honor the contract. As a result, Garrett launched his own enterprise, Garrett & Company, a wine-distributing firm at Littleton. The dispute, nonetheless, involved a lawsuit in the superior court of Halifax County. In 1891, the parties concerned, C.W. Garrett & Company (made up of Lucy W. Garrett, Charles W. Garrett's other daughter, Henry Spooner Harrison and Mary Garrett Harrison) and Paul Garrett, negotiated an end to the conflict. Both parties agreed to accept arbitration in order "to avoid the delay, expense, and unfriendly relations which necessarily attend litigation." According to the legal proceedings, the arbitrators addressed three issues:

> *First, What amount does Paul Garrett, trading as Garrett & CO., owe to C.W. Garrett & Co. for wines, brandies, and other merchandise sold and delivered:*
>
> *Second, What amount, if any, is Paul Garrett entitled to recover of C.W. Garrett and Co. for the alleged breach or breaches of contract which existed between said parties.*
>
> *Third, It is agreed that…in the event the arbitrators shall conclude that either party has been damaged by the course of the other, they shall decide the amount of such damages.*

In the end, the arbitrators decided that Garret & Co. should pay C.W. Garrett & Company $2,269.51 and that C.W. Garrett and Company should pay Garrett & Co. $117.00 on account and $67.61 for sour wine and other matters. Unfortunately, arbitration did not end the legal disagreements between C.W. Garrett & Company and Garrett & Co. The former sued the latter for damages "suffered by reason of the defendant trading" under the Garrett name, and Paul Garrett also sued to vacate the arbitration award. Yet, eventually, the Garretts resolved their differences.

Later, Paul Garrett moved his operation to Chockoyotte, near Weldon, and built a wine-processing plant there. In time, he leased the vineyard and winery at Medoc from his cousin, Mary Garrett Harrison, wife of Henry Spooner Harrison. Medoc, however, was not destined to be Paul Garrett's path to national success. The old winery burned down in 1906 and was not rebuilt.

Between 1900 and 1916, Paul Garrett built a national wine operation, bottling and selling primarily scuppernong wine. The Weldon location was logistically better than Medoc. The plant could ship its wine by rail or over the Weldon Canal. Savvy business sense and cutting-edge marketing guided Garrett's success. The naming of his signature wine illustrates his business acumen.

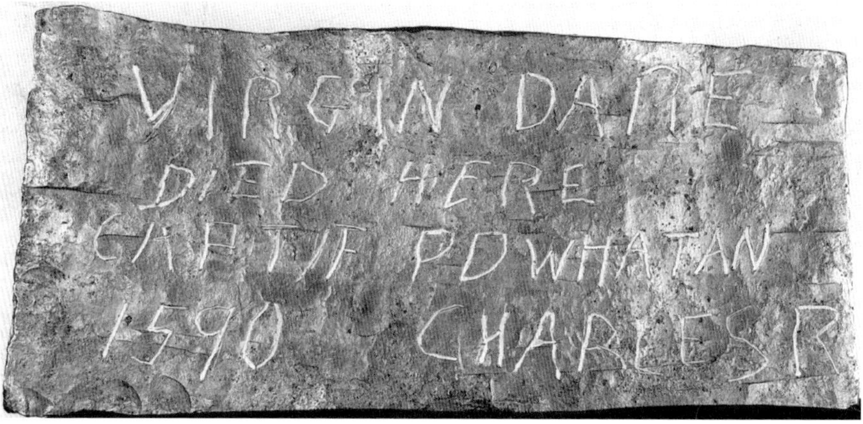

Virginia Dare marker. Charles Garrett named his bestselling vine "Virginia Dare" in honor of the first English child born in North America. *Courtesy of the Library of Congress.*

In a brilliant tribute to the story of the Lost Colony and the Mother Vine, Garrett named his new wine Virginia Dare. Virginia Dare, born August 18, 1587, was the first English child born in North America. Her parents, Eleanor and Ananias Dare, were members of the English colony on Roanoke Island. Her grandfather, John White, was governor of the colony. The fate of the colonists continues to intrigue new generations. The name "Virginia Dare," then, tapped into a deep pool of romance and myth. Garrett's Virginia Dare wine, which came in red and white varieties, became the bestselling wine in the United States and, for decades, dominated the pre-Prohibition wine market.

In 1904, Paul Garrett's champagne garnered the grand prize for sparkling wines at the Louisiana Purchase Exposition. By 1905, Garrett had five wineries in North Carolina and was known nationally. Garrett packaged his wines in either glass bottles or wooden barrels. The glass bottles, designed by a Weldon artist, included a Confederate flag in the Garrett label. In 1905, Garrett sold a variety of wines, including White Imperial Scuppernong, port, sherry, Catawba and cognac. Always alert to market options, C.W. Garrett & Company also produced a line of sacramental wines. Garrett sold so much wine that North Carolina vineyards could not supply enough grapes. Increasingly, he purchased grape juice from California to produce his wines. As time went by, there was less and less scuppernong in Garrett's wines.

Yet there were storm clouds on the horizon. As adherents of the Temperance movement spread, more and more North Carolina counties went dry. As a result, Garrett had to relocate his wine-production facilities. By 1909, when the State of North Carolina banned alcohol production and consumption, Garrett had moved his operation across the line to Berkley, near Norfolk, Virginia. Later, he transferred his headquarters to New York. From there, the Garrett enterprise diversified and spread across the nation. Garrett continued to innovate as he adapted, marketed his wines and diversified his business plan. As a result, he was not only North Carolina's best-known winemaker but also a national force in the wine industry.

In addition to the Garretts at Medoc, there was a colony of French grape growers in Ridgeway, Warren County, North Carolina. In 1870, Eugene Morel, a pupil of the French viticulturalist Jules Guyot, attempted to plant European grapes near Ridgeway. Other French settlers also moved to the area. The 1880 census information suggests that some of these farmers may have been French Canadians or, at least, Frenchmen who lived in Canada before immigrating to the United States. Ultimately, the experiment was not successful.

However, in 1873, Jules Emile Planchon (1823–1882), a noted French botanist on a tour of the United States, visited the French grape growers. Planchon was director of the Superior School of Pharmacy at the University of Montpellier (Montpellier, Hercault). The French Ministry of Agriculture had sent him to tour the phylloxera-resistant vineyards of the eastern United States. North Carolina muscadines were resistant to the phylloxera, a green aphid or louse that had been destroying French vineyards. Phylloxera began attacking French vineyards about 1860. Planchon's tour and the work of Charles Valentine Ripley (1843–1898), first entomologist for Missouri, were critical in averting this late nineteenth-century viticultural catastrophe in France. Today, few Americans are aware of the critical role American vines played in saving the French wine industry.

Planchon, a friend and colleague of Ripley's, devoted a happy (by all accounts) week to touring and studying American and French grapes in North Carolina. Planchon favorably compared the Ridgeway area with Provence in France. The North Carolina Agricultural Society in

Raleigh also sent a delegation to meet with the visiting dignitary. They invited Planchon to attend the North Carolina State Fair. On September 13, Planchon left North Carolina and continued his vineyard tour in Cincinnati, Ohio. However, some believe that the Frenchman accepted the invitation and returned to Raleigh for the state fair in October 1873.

Not all North Carolina wine was legally produced. In May 1878, a raid in Granville County, as reported to the commissioner of internal revenue in Washington, D.C., netted, in addition to illegal stills and untaxed beer, "fifty gallons of low wines."

Also in the 1870s, Louis Froelich (1817–1873) turned to farming and winemaking. During the Civil War, Froelich, who was born in Bavaria, and a partner opened the CSA Arms Factory in Wilmington. He later moved the concern to Kenansville, Duplin County, and manufactured swords, sabers, buttons, guns and other equipment for the Confederate military. In 1863, a Federal raiding party damaged his plant, but Froelich persisted with his arms manufacture. After the war ended, Froelich pursued other interests, including farming and winemaking. As a vintner, Froelich produced acclaimed scuppernong wines. He also actively promoted the cultivation of muscadine grapes. Among Froelich's grapes were Froelich's White Seedling, Pamlico, Beaufort and Sugar. Before his death in 1873, Froelich and his family moved again to Halifax County.

Tokay

Colonel Wharton Jackson Green (1831–1910), a native of Florida, was an attorney, planter, Confederate veteran and member of the United States Congress. He attended the United States Military Academy at West Point, was commissioned a lieutenant colonel during the Civil War and was wounded and captured during the Battle of Gettysburg. Green settled at Esmeralda in Warren County. However, in 1879, Green acquired the Tokay Vineyard about three and a half miles north of Fayetteville, Cumberland County, North Carolina and relocated there. The Tokay Vineyard, according to his memoirs *Reflections and Recollections* (1908), was "said to be the largest one this side of the Rocky Mountains." Tokay

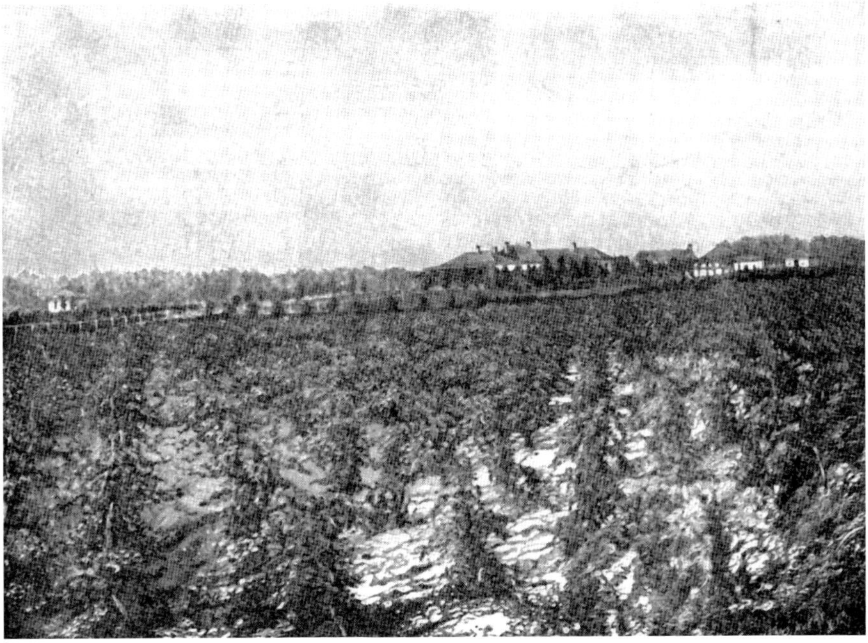

Tokay Vineyard from the southwest, Fayetteville, North Carolina. In 1883, Tokay Vineyard, operated by one-time United States congressman Wharton J. Green, covered two hundred acres. *Wharton J. Green,* Tokay Vineyard near Fayetteville, N.C. with essay on grape-culture.

at that time had approximately sixty acres of muscadines, principally, scuppernong, Flowers and Mish. The wines were widely available, as in 1878, Reed's Sons advertised in the *Washington Post* red and white sweet scuppernong wines and a "red dry seedling wine" from the "celebrated Tokay Vineyard" of North Carolina. Tokay wines in 1878 were "highly recommended for their purity and medicinal qualities."

According to the *Times Dispatch* of Richmond (1903), Henry L. Myrover and W.T. Horne, Fayetteville merchants, originally planted Tokay "about fifty years ago," (circa 1853). Myrover and Horne allegedly planted the vineyard "not so much for money as for a recreative [*sic*] occupation, and to vindicate their belief that good, pure wine, as diminishing the demand for whiskey and brandy, was promotive [*sic*] of temperance." In 1860, Myrover was a prosperous wholesale merchant who lived with his wife, Mary, and their large family in Fayetteville. In addition, Myrover was president pro tem of the North Carolina Constitutional Convention of 1865.

From these sixty acres of grapes, Green produced between fifteen and twenty-five thousand gallons of wine per year. By 1892, he had one hundred vines and cultivated, in addition to the *rotundifolia* vines, several varieties of *Vitis labrusca*, including the Norton, Cynthia, Herman, Martha, Champion and Concord, as well as Ives, Delaware and Cottage grapes. Superintendent McBuie managed Tokay Vineyard for Green. An 1883 visitor to Tokay described Green's home as "being beautifully located, and handsome in its surroundings, nestled as it is in the midst of the beautiful Tokay Vineyard, surrounded by almost innumerable trellised grape-vines of equal height." The estate also included fruit trees and fishponds.

On November 27, 1880, Green left his Warren County home and moved to Tokay, where he lived until his death. The vineyard was a good investment. As Green noted, Tokay "has been a source of solace [his first wife died in 1883] to me, saying nothing of profit." Green also served as a Democrat in the Forty-eighth (1883–84) and Forty-ninth (1885–86) U.S. Congresses. He then returned to Tokay, remarried and lived near his vineyard until his death in 1910. Green was also the first president of the Society of Confederate Soldiers & Sailors in North Carolina. In 1893, he served as chair of the State Board of Agriculture of North Carolina.

According to published accounts, Green produced superior wines. At the Atlanta Exposition (probably the International Cotton Exposition held at Oglethorpe Park, Atlanta, in 1881), Colonel W.J. Green's wines "won the first premium of $15" for "the best collection of sparkling wines." In addition, his non-sparkling wines also earned a second premium. He compared his wines to Madeira, sherry, California Mission and Spanish wines. According to Green, his wines had a "dual" character, as there were "red and white wines made from the grapes that flourish in the States further north" such as the "Norton and Cynthiana, the Delaware and Martha" and, on the other hand, the "Scuppernong, and its children the Flowers and Meisch [Mish]," which were "peculiar to the States south of the thirty-sixth parallel of north latitude." In other words, Green was producing wines from a variety of both *Vitis rotundifolia* and *Vitis labrusca* native grapes. Concerning the scuppernong, Green wrote, "It was said that the first wine made on this continent was made from this vine."

His clients frequently wrote testimonials, some of which Green included in his *Tokay Vineyard, near Fayetteville, N.C. with essay on grape-culture*, published in 1883. A reporter for the *Raleigh News and Observer* wrote that Green's "Claret and Scuppernong are superior to any I have ever tasted." Green's wines apparently were well received by officers of the United States Army as well as professors at West Point. For example, Professor J.B. Wheeler of the United States Military Academy commended Green's Rhine wine and recommended his "Sweet Meisch [Mish]" as a "wholesome and healthy drink" free from the "adulterations and impurities of the foreign stuffs which characterize the importations of the day from abroad." Even in 1883, patriotism was a factor in marketing campaigns. Charles Levi Woodbury of Boston liked Green's claret for "its richness of aroma, and fruity quality" and also found the scuppernong "a capital ladies' wine."

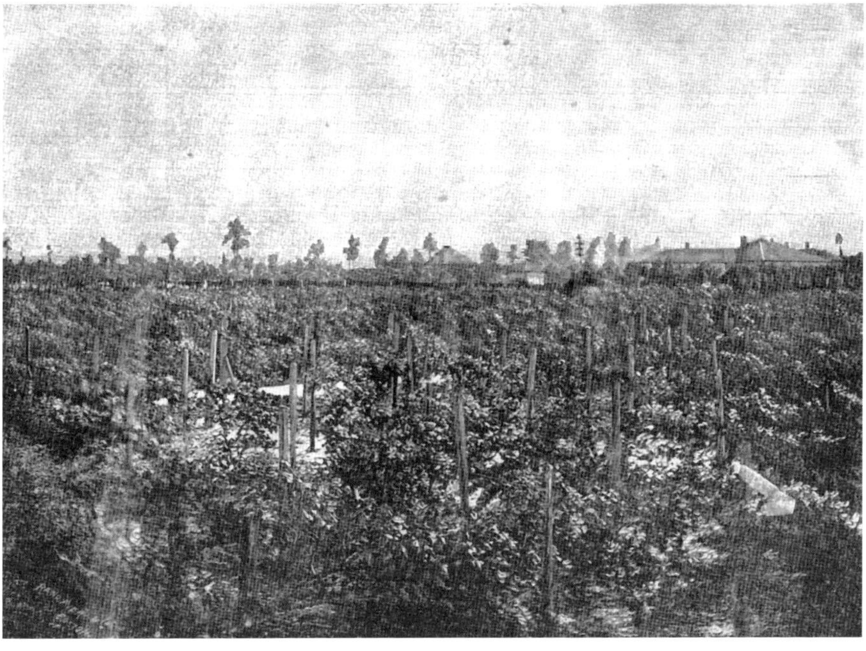

Tokay Vineyard from the northeast, Fayetteville, North Carolina. Tokay wines were widely distributed and highly acclaimed. One satisfied customer wrote, "The claret…is as good as any I ever drank," while another (a professor at West Point) commented that Tokay produced "a native wine" that was "equal, if not superior, to the finest Rhine wines." Professor Wheeler continued, "I had no idea such wine was made in this country." *Wharton J. Green,* Tokay Vineyard near Fayetteville, N.C. with essay on grape-culture.

An army doctor—W.H. Gardner of Fort Davis, Texas—similarly praised the claret and commended the "Sweet Meisch" as a "universal favorite with ladies " and invalids. Tokay wines also won prizes at North Carolina state and county fairs.

Perhaps the following contribution from a northern journal reproduced in the 1893 *Handbook of North Carolina* sums up Green's wines at Tokay:

> *In general characteristics they resemble the Spanish and Madeira wines, and the Sweet White is not unlike the California Mission, though much more delicate in bouquet, and, when given proper age, approaches the closest to a fine Madeira of any wine yet produced in this country. This wine will constitute a good basis for a sherry wine when made with that view, and we have seen some samples of such from these vineyards which strongly resemble Old Brown Sherry, and would be credit to any gentleman's sideboard and private cellar. Other samples, again, made from the Flowers, a black Scuppernong seedling—a dry wine—resembles certain red wines of Hungary already highly esteemed in this country and, as a sweet wine, bears a close relation in character to Spanish Red.*

Green ceased producing his much applauded wines by 1903. Colonel Green then leased the vineyard to Garrett & Company of Weldon, North Carolina.

Another component of the eastern North Carolina wine story was Sol Bear & Co. Fleeing religious persecution, in 1852 the Bear brothers, Solomon and Samuel, emigrated from Bavaria to Wilmington, New Hanover County. Together, the brothers established a successful mercantile business in Wilmington. Solomon Bear was also civic-minded, involved with the Jewish community and a Civil War veteran. In 1860, Bear had a personal estate of $18,000, and in 1880, Solomon Bear, his wife, Henrietta, and their six children (Miriam, Jacob, Teresa, Rebecca, Isadore and Pauline) lived in Wilmington.

In 1882, the Bear brothers parted company, and Solomon Bear purchased Castle Hayne Vineyard (incorporated between 1867 and 1872 in Virginia). The new enterprise, Sol Bear & Co., retailed spirits and bottled wine. Bear & Co. produced scuppernong and blackberry wines

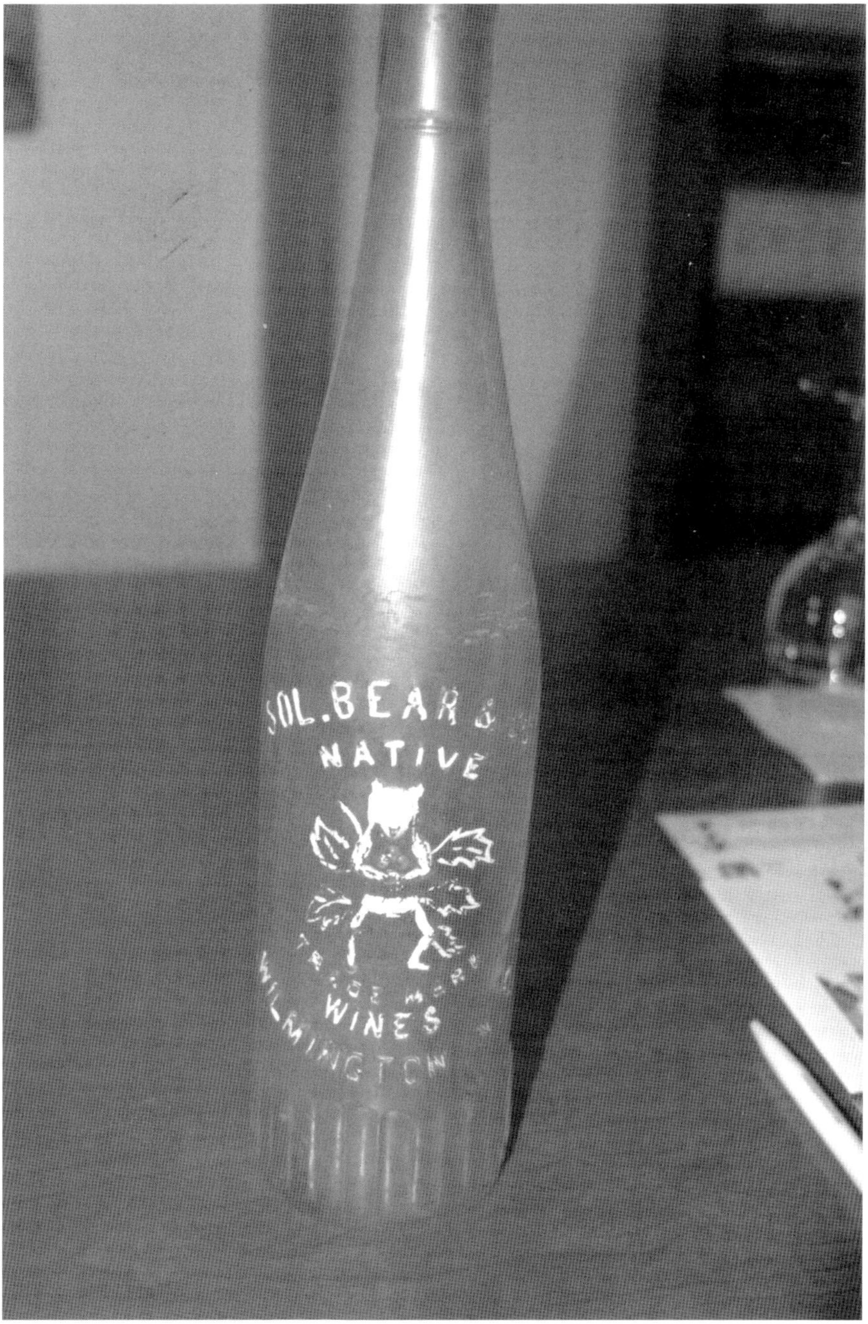

Sol Bear & Company Native Wines bottle, Wilmington, North Carolina. Immigrant Solomon Bear and his brother operated a successful mercantile business in Wilmington before Bear opened Sol Bear & Company in 1882. Bear & Company sold wine and liquor. *Courtesy of David G. Fussell Sr.*

and a local version of port called the "Carolina Belle." The business prospered, and at one time, Bear was also president of the North Carolina Liquor Dealers, Distillers, and Grape Growers Association. After Bear's death in 1904, his sons Isadore and Fred Bear (a later son) managed the company and constructed new facilities on the corner of Front and Marsteller Streets in Wilmington. However, when the State of North Carolina went dry in 1909, business declined, although, through a legal loophole that permitted North Carolinians to purchase alcohol for medical purposes, the company pursued that niche market and stayed in business until 1914. At that time, Isadore and Fred Bear sold the winery to Paul Garrett. Garrett had earlier left North Carolina as Prohibition claimed more and more North Carolina counties.

In addition, there were several other important late nineteenth-century eastern North Carolina vineyards. Cumberland and Moore Counties were centers of grape growing. Generally, the growers preferred Niagara, Delaware, Concord and scuppernong grapes. The Garrett plant at Aberdeen, as discussed elsewhere, paid handsomely for scuppernong

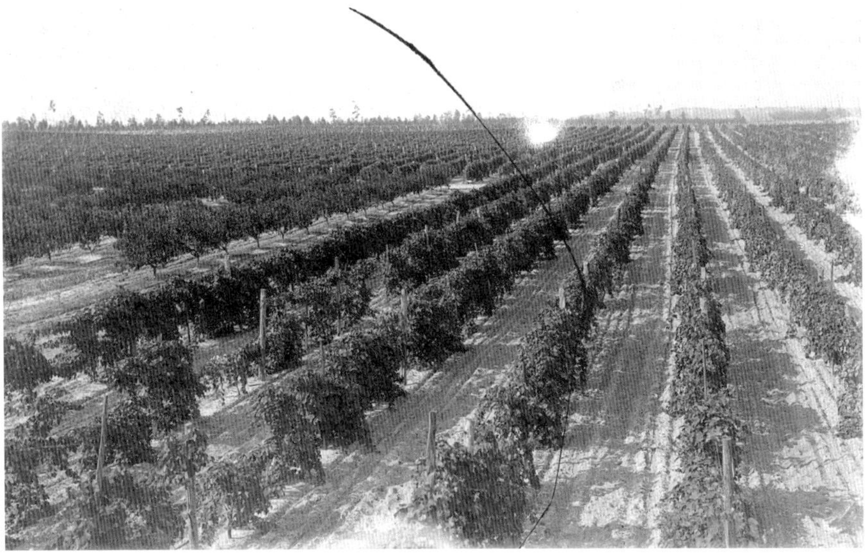

Niagara Vineyard, Moore County, North Carolina, circa 1900. Niagara Vineyard cultivated 250 acres of grapes, primarily for shipment. *Courtesy of the North Carolina State Archives.*

grapes. T.S. Lutterloh had been raising grapes in Cumberland County since the 1850s. In 1860, Thomas S. Lutterloh was a forty-four-year-old turpentine merchant living near Fayetteville.

One of these vineyards, the Bordeaux Vineyard, stood three or four miles west of Fayetteville in Cumberland County. James M. Pearce, the owner, planted the Bordeaux Vineyard primarily with scuppernong grapes. In 1893, the vineyard had five hundred bearing scuppernong vines plus other grapes as well. Pearce shipped his Bordeaux grapes principally to western North Carolina. In 1900, Pearce was a forty-eight-year-old farmer in Ray's Precinct, Cumberland County.

By 1896, there was another vineyard near Fayetteville named Happy Valley Vineyard. George W. Lawrence was the owner of Happy Valley Vineyard, which had about four hundred acres of mainly scuppernong grapes. One vine covered a space seventy-two by eighty-four feet. Happy Valley produced thirty or more bushels of grapes annually. Lawrence advocated scuppernongs as table grapes and for "the finest bright yellow wines." In 1900, Lawrence was a seventy-two-year-old farmer in Cross Creek Township, Cumberland County, and the census enumerator.

Captain S.W. Noble owned one of the "large and valuable" vineyards planted near Wilmington, New Hanover County, and in 1899, Cleveland County had Cleveland Vineyard Company at Shelby. Also, there was a large vineyard, Southern Pines Grape Nurseries, managed in 1913 by R.C. Cool, of several hundred acres at Southern Pines in Moore County. Vineyards in the Southern Pines area—such as Southern Pines and the Niagara Vineyard Company—usually did not make wine, as it was too profitable to ship their grapes. In 1896, there was also a large scuppernong vineyard near Whiteville (once known as Vineland) in Columbus County, and the Whiteville Wine Company exhibited samples of its wine in St. Louis for the Louisiana Purchase Exposition (also known as the 1904 St. Louis World's Fair).

In the late nineteenth century, Wake County had a grape growers' association and a farmers' institute. Identified grape growers in Wake, Halifax and Warren Counties included B.P. Williamson, J. Heck, H. Mahler, George Shellem (an English-born farmer), Batchelor & Womble, Harrison Bilyeu (a Pennsylvania-born nurseryman), C.B. Edwards, V. Royster, B.G. Cowper, M.A. Parker, J.B. Burwell, Dr. Lewis, Davis &

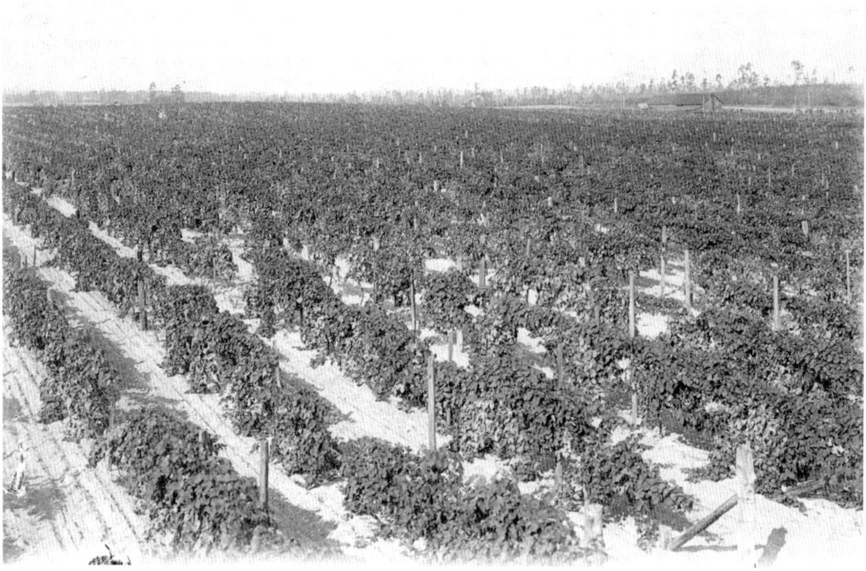

Grape vineyard, Southern Pines. This photograph may show vines from the Southern Pines Grape Nurseries. In 1913, R.C. Cool, manager of the nursery, published *The Scuppernong Grape: Its Growth & Care under Vineyard Conditions. Courtesy of the North Carolina State Archives.*

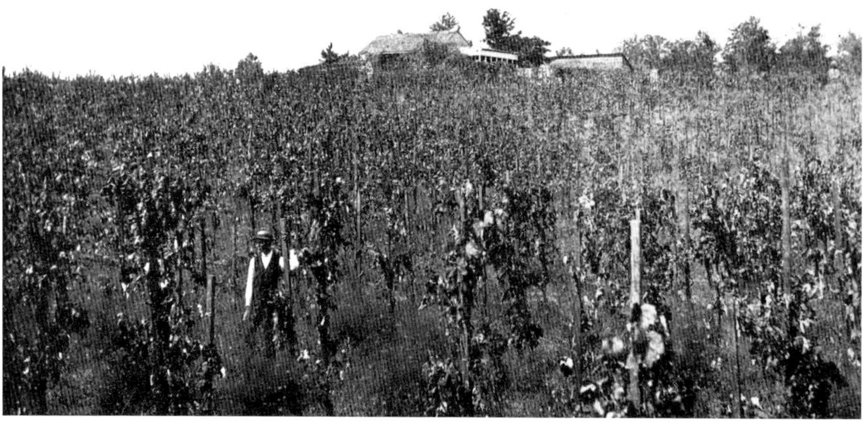

Mahler's Vineyard near Raleigh, North Carolina. In the 1890s, H. Mahler owned a vineyard near Raleigh in Wake County. *Courtesy of the North Carolina State Archives.*

Bradshaw, Philip H. Andrews, Dr. Fuller, C.D. Upchurch, George Cole, Frederick Watson, J.S. Otho Wilson (Raleigh Nurseries at Swift Creek), Robert Strong, the Whiting Brothers, W.H. Pace, Ferguson, Jesse Jones, James Moore, John Robinson (commissioner of agriculture) and Judge Walter Clark. In Wake County, the favorite grape varieties were Concord and Ives. Most of these vineyards shipped their produce to northern markets. A few such, as H. Mahler, converted their grapes into wine. Others did so only when a glut of grapes reduced the external market. In 1893, grape growers near Raleigh had 359 acres in cultivation with an annual yield of 89,750 baskets. Also, smaller vineyards were popular in the western parts of the state.

Western North Carolina

In the post–Civil War years, grape cultivation and wine production prospered in Polk and Buncombe Counties of western North Carolina. By the 1890s, Polk County (especially Tryon Township) was a center of grape and wine production in western North Carolina. In 1886, George Edward Morton, founder of Tryon's first newspaper, the *Tryon Bee*, moved from Cincinnati, Ohio, to Tryon. Morton grew the Tryon grape Valhalla on his aptly named Valhalla Fruit Farm in the Pacolet River Valley. According to the 1900 census, George A. Morton, also a native of Ohio, and a horticulturalist, lived with his wife and daughter in Tryon Township. In 1893, William T. Lindsey, owner of a nearby thirty-acre vineyard on Warrior Mountain, sent grapes for exhibit to the New York State Fair. In 1900, the census listed Lindsey, a native of Kentucky, as a fruit grower. R. Henry Scadin (1861–1921) was another transplant. Scadin, a fruit grower and photographer who had lived in Michigan and Vermont, moved to western North Carolina in 1889. In 1896, Polk County, North Carolina, had its first commercial shipment of grapes. In 1884, the scuppernong grape received international attention. In an effort to cultivate American fruit in China, scuppernong vines and fruit trees were shipped from Raleigh to Shanghai.

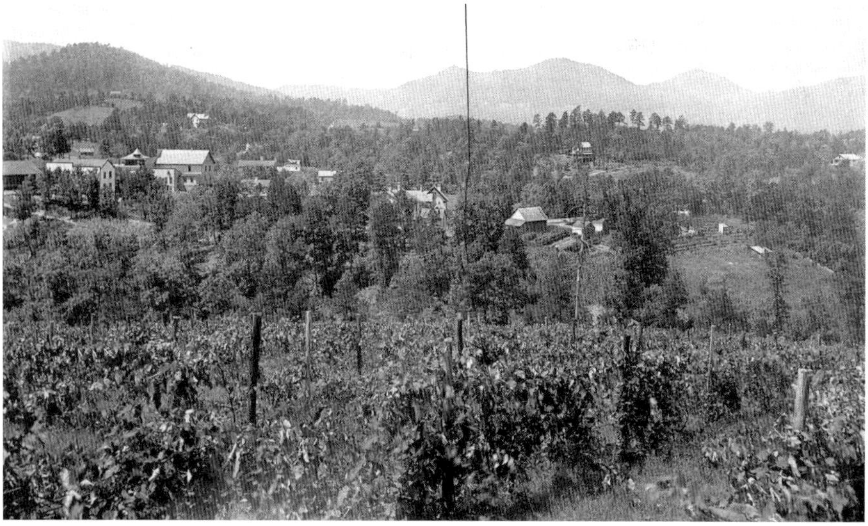

Western North Carolina vineyard. The western North Carolina counties of Buncombe and Polk were significant grape-growing and wine-producing areas before Prohibition. Today, wineries once again flourish in these counties. As Edward Antill wrote in 1771, "But if it be a mountain, with a rich soil, it will be best of all; for the higher the vineyard, the richer the Wine." *Courtesy of the North Carolina State Archives.*

In the 1890s, Union general Ulysses Doubleday (1824–1893), brother of General Abner Doubleday of baseball fame, established a vineyard on Laurel Avenue on the east side of Piney Mountain. During the Civil War, Ulysses Doubleday served as commander of the Sixth New York Heavy Artillery, lieutenant colonel of the Third United States Colored Troops and colonel and commander of the Forty-fifth United States Colored Troops. Near Tryon, Doubleday, once a member of the New York Stock Exchange, cultivated white Niagara and red Delaware grapes. Doubleday's venture, according to Tryon's 125th anniversary proclamation, initiated "an era of highly successful viticulture on Tryon area slopes."

After the general died, his son Harold Doubleday (1867–1925) and Sidney Lanier Jr., the son of Sidney Lanier, operated the vineyard. Sidney Lanier (1842–1881) the father was a southern poet, composer and professor at Johns Hopkins University in Baltimore, Maryland. Lanier, a native of Georgia, enlisted in the Macon (Georgia) Volunteers. Captured in 1864, the poet contracted tuberculosis in prison. One of

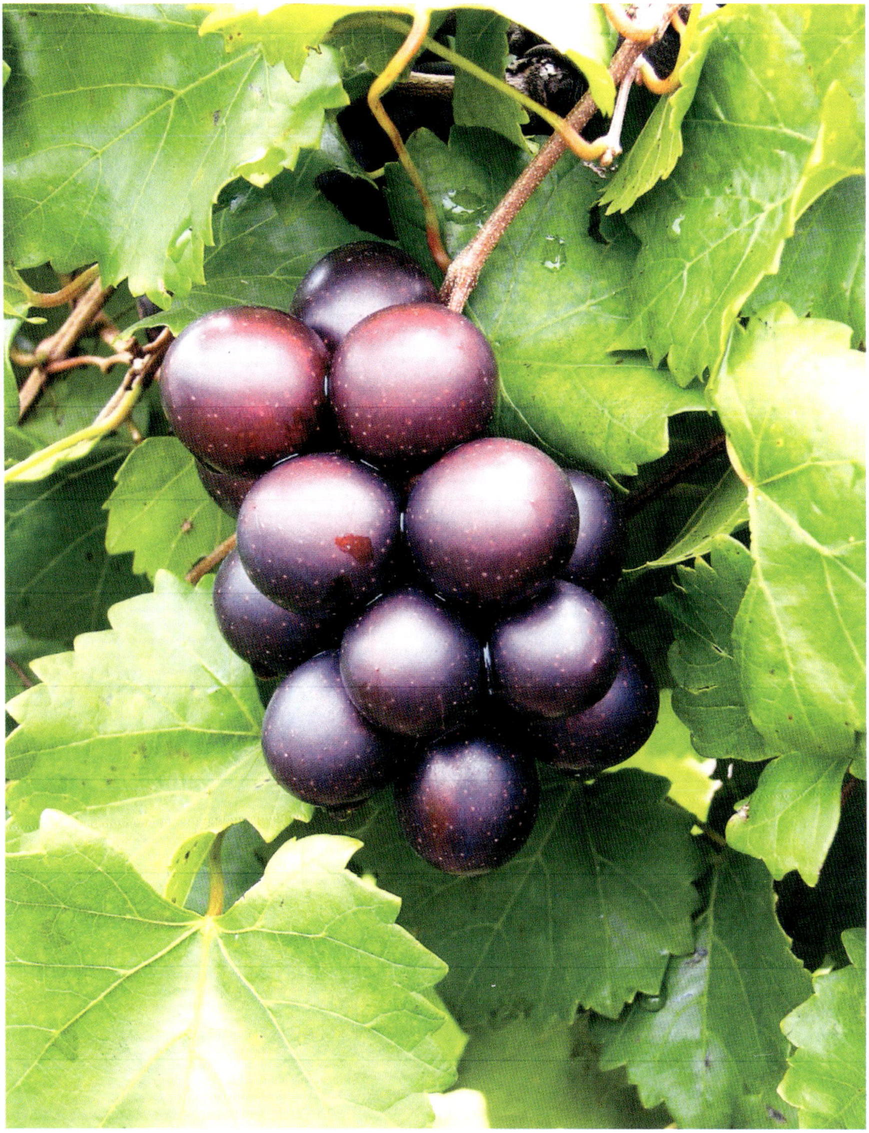

Alachua muscadine grapes. "Like grandfather's scuppernong grapes…too sweet to resist." From "Pamlico River" by James Applewhite. Muscadine grapes make great jellies and jams and are a favorite freshly picked from the vine. Alachua is a muscadine cultivar from Florida. *Courtesy of the North Carolina Wine & Grape Council.*

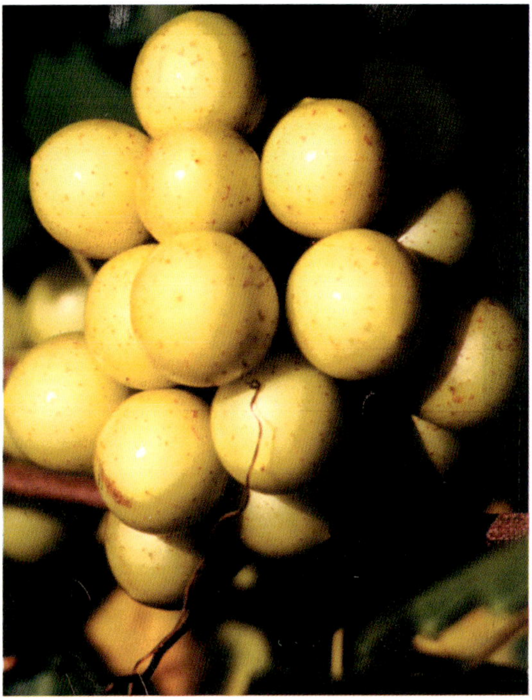

Bronze muscadine grapes. The color of muscadine grapes varies greatly. Among the varieties are red, deep purple, bronze, green and white. In 1920, there were thirty-five to forty named varieties, but fewer than ten varieties were cultivated commercially. Popular cultivars now are Carlos, Magnolia, Sterling, Nesbitt and Noble. *Courtesy of the North Carolina Wine & Grape Council.*

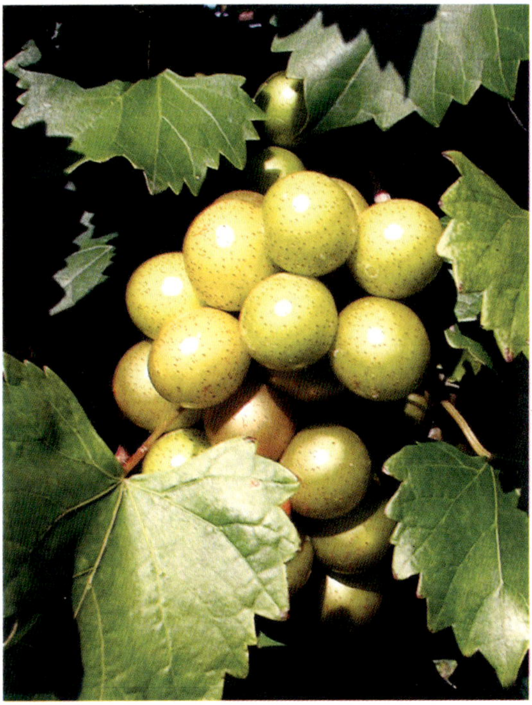

Muscadine grapes. The fruit of the scuppernong is, according to D.H. Jacques, large, up to an inch in diameter, sweet and "pale green when fully ripe, and dotted with brown." Muscadine grapes do not grow in bunches. *Courtesy of the North Carolina Wine & Grape Council.*

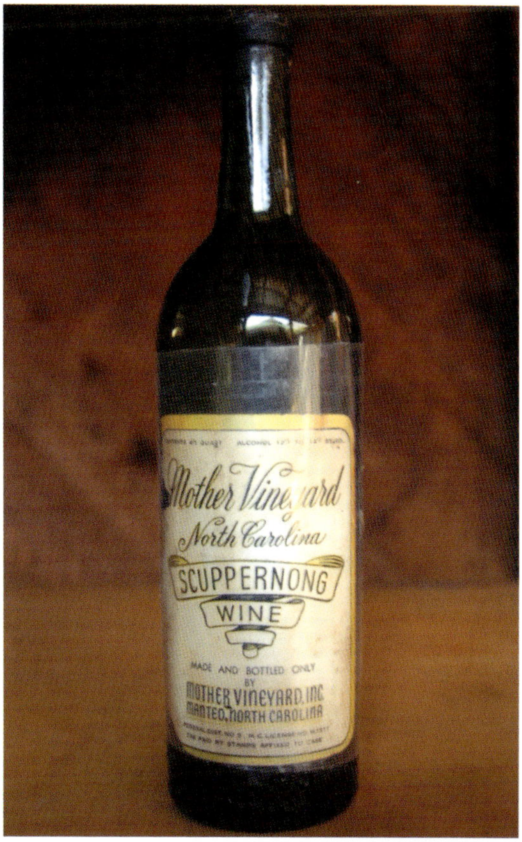

Right: Bottle of Mother Vineyard scuppernong. The label "Mother Vineyard" originally indicated wine produced from the grapes of the Old Mother Vineyard on Roanoke Island. *Photograph by Terry L. Helsley, courtesy of David G. Fussell Sr.*

Below: Mother Vineyard, Manteo. This vineyard is part of the romantic history that surrounds the scuppernong grape. W.B. Fearing Co., Manteo, North Carolina, Curteich-Chicago "C.T. Art-Colortone." *Courtesy of the author.*

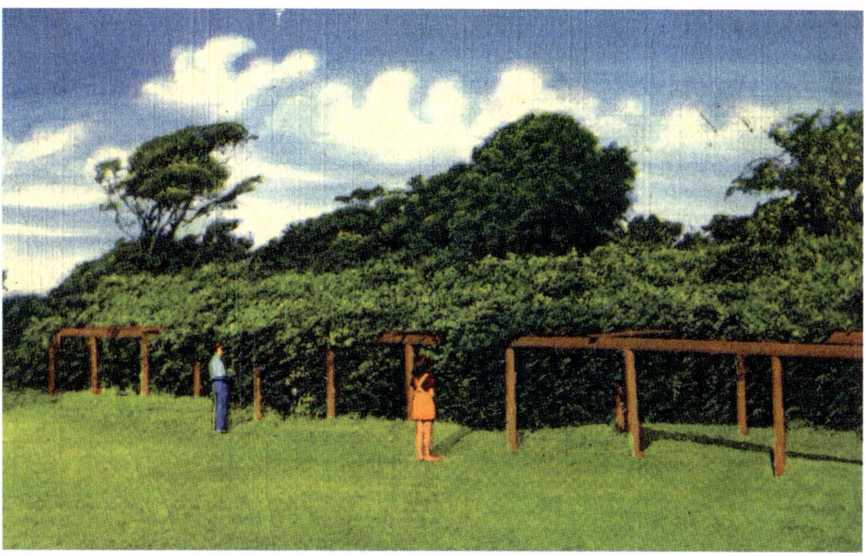

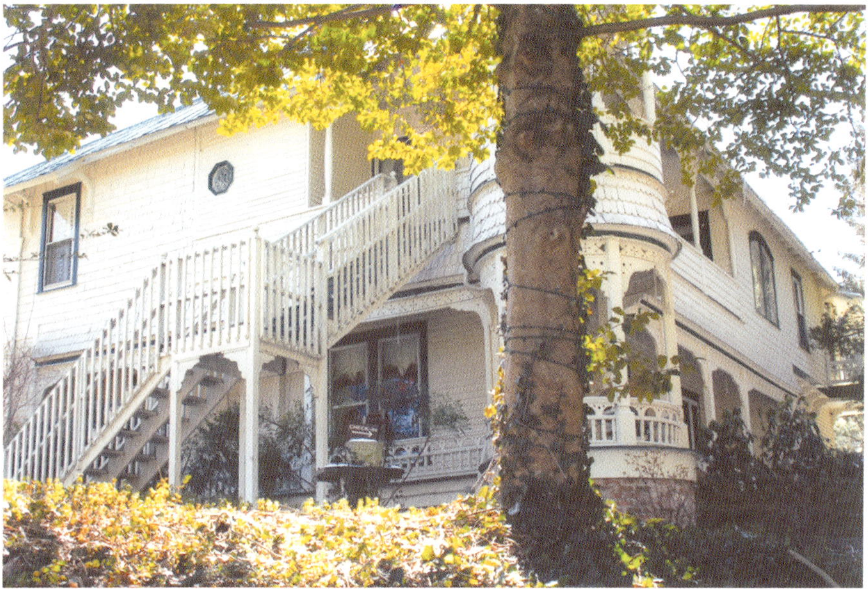

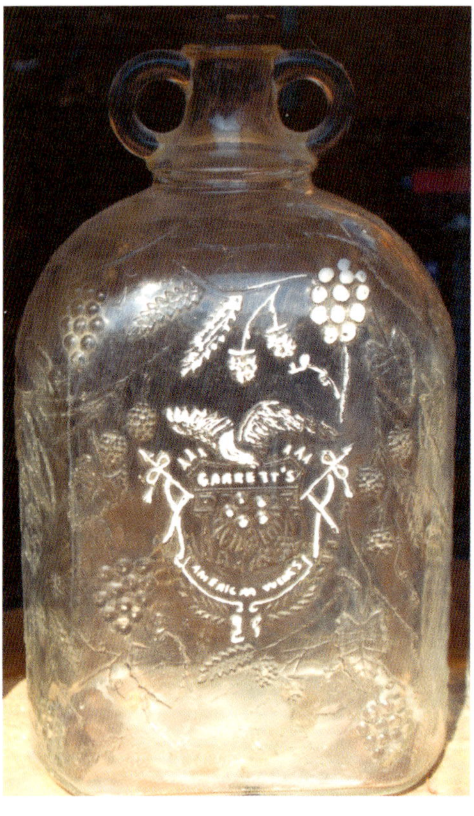

Above: Engadine house. John K. Hoyt's mansion at Engadine Vineyards near Asheville was unusual in 1885. The house had running water, central heat and electricity (from a generator). *Courtesy of Terry L. Helsley.*

Left: Garrett two-handled bottle. Garrett & Company advertised that its wines were "always sold" in its "own bottles" with its "own name and trade-mark blown in the glass." *Photograph by Terry L. Helsley, courtesy of David G. Fussell Sr.*

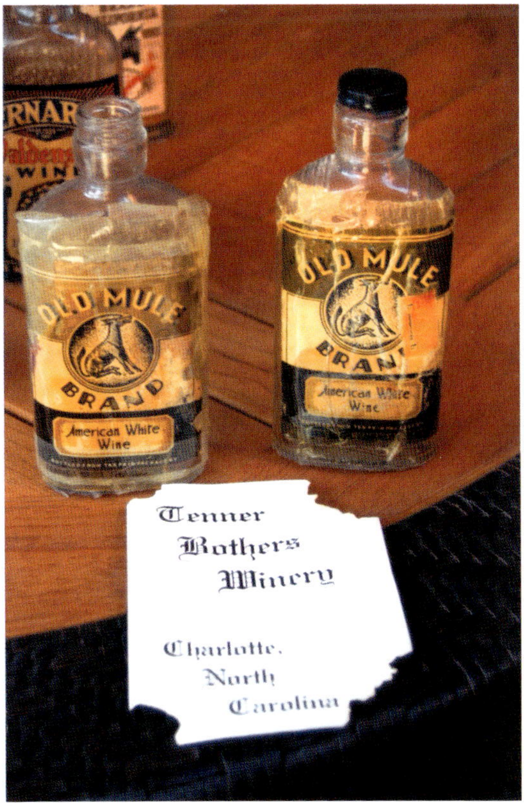

Right: Bottles, Tenner Brothers, Charlotte. In 1935, Sol, Albert and Leon Lukie Tenner opened a winery and restaurant in Charlotte. Later, they planted a large vineyard near Patrick, South Carolina, and moved Tenner Brothers winery across the state line to South Carolina. *Photograph by Terry L. Helsley, courtesy of David G. Fussell Sr.*

Below: Duplin Wine Cellars sign. In the 1970s, two brothers—David and Dan Fussell—and several other grape growers organized Duplin Wine Cellars at Rose Hill. *Photograph by Terry L. Helsley, courtesy of David G. Fussell Sr.*

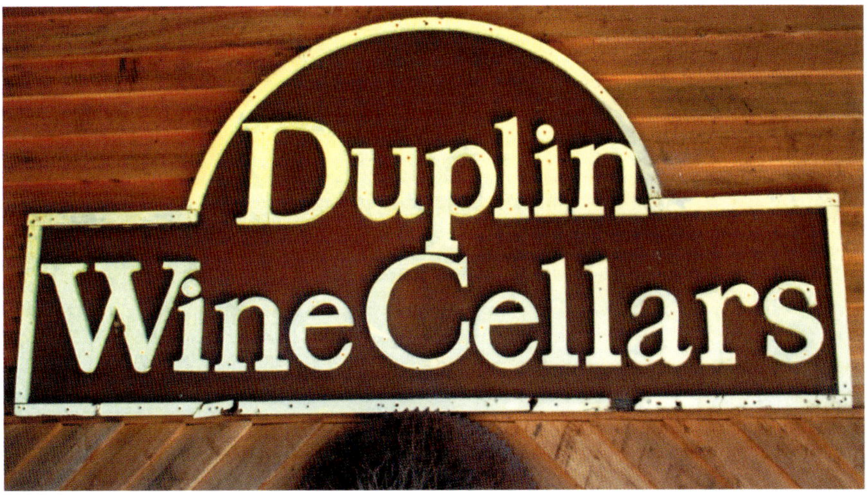

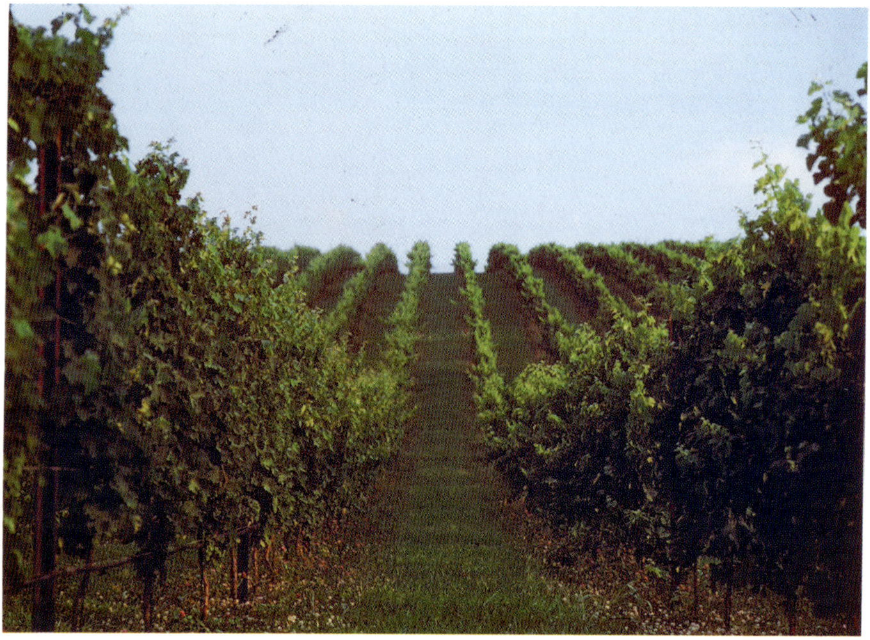

Vineyard Rows, Shelton Vineyard, Dobson. The eye-pleasing symmetrical rows of grapevines are part of the charm of North Carolina's vineyards. *Courtesy of George A. Denka.*

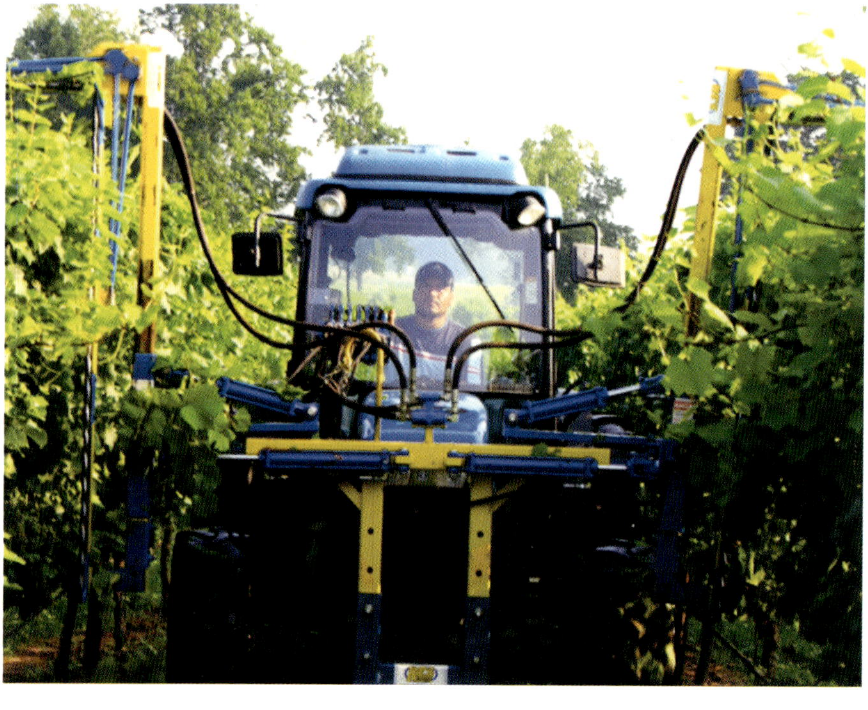

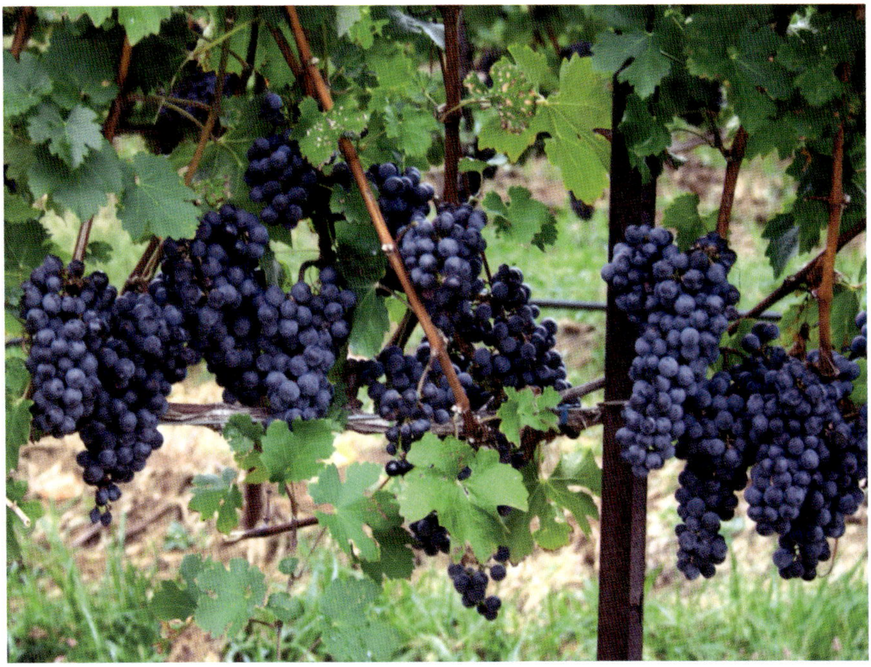

Above: Cabernet grapes hanging. North Carolina vineyards are prized not only for their grapes and wines but also for their aesthetic value in preserving an agricultural landscape. *Courtesy of George A. Denka*.

Right: Viognier grapes. A European grape, the viognier grows in the Upper Rhone Valley of France. Since the 1980s, American vintners have cultivated the grape, which produces, according to a representative of Rockhouse Winery, a "light and crisp wine" that's perfect for picnics. *Courtesy of George A. Denka*.

Opposite bottom: Tractor hedging. Modern vineyards meld state-of-the art technology with old-fashioned care to produce fine wine. *Courtesy of George A. Denka.*

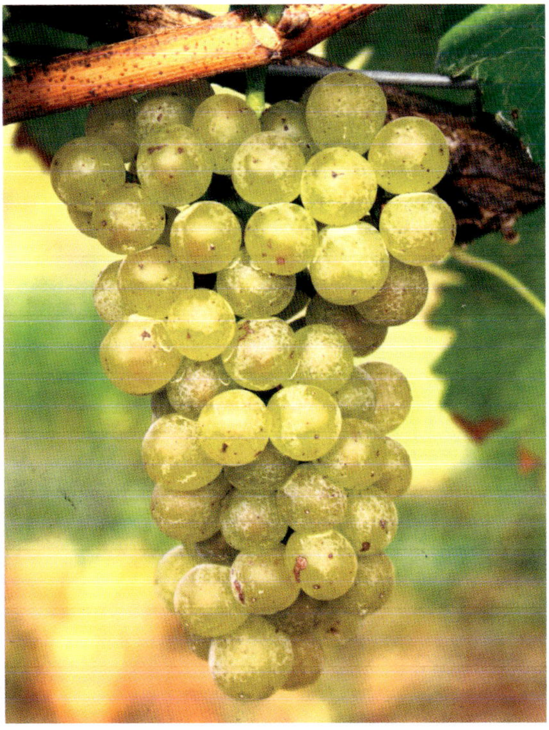

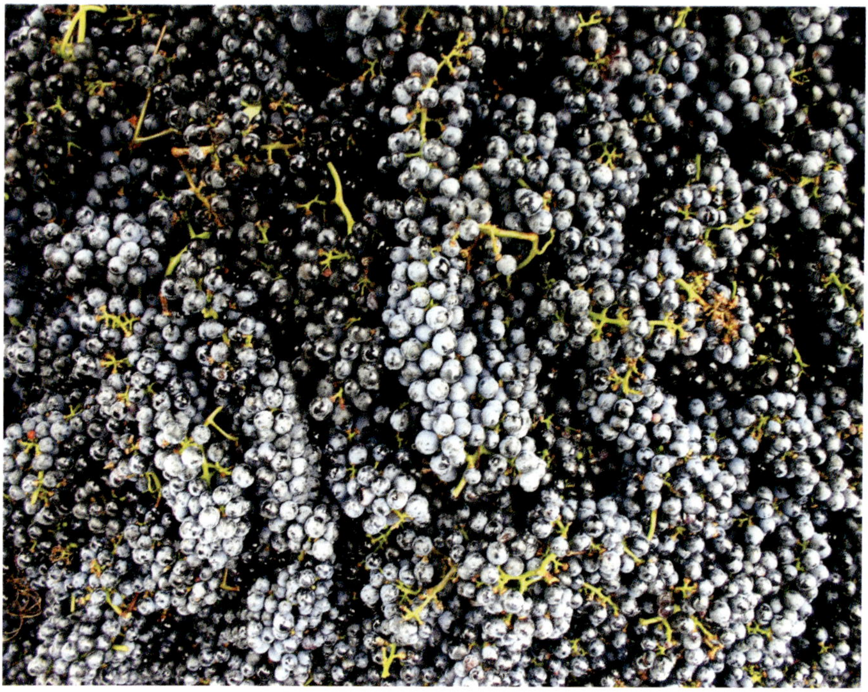

Cabernet sauvignon grapes. Cabernet sauvignon grapes are the dominant grapes of the Bordeaux region of France. This big red-wine *vinifera* grape is one of the most widely planted grapes in the world. *Courtesy of George A. Denka.*

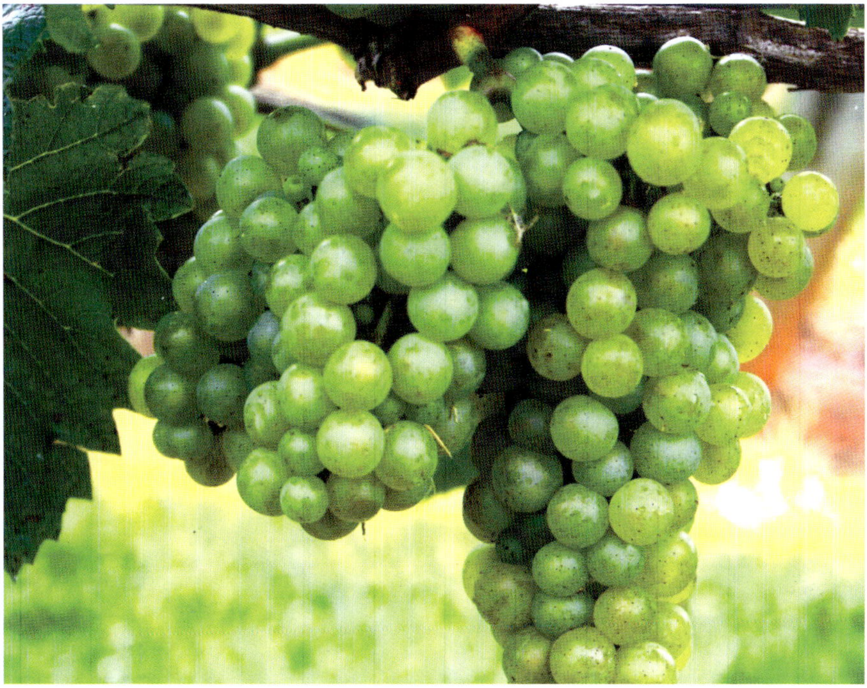

Chardonnay grapes. Vintners use the green-skinned chardonnay grapes to produce a popular white wine. This *vinifera* grape is probably the most widely grown grape in the world. With its short growing cycle, it is an adaptable and popular grape. *Courtesy of George A. Denka*.

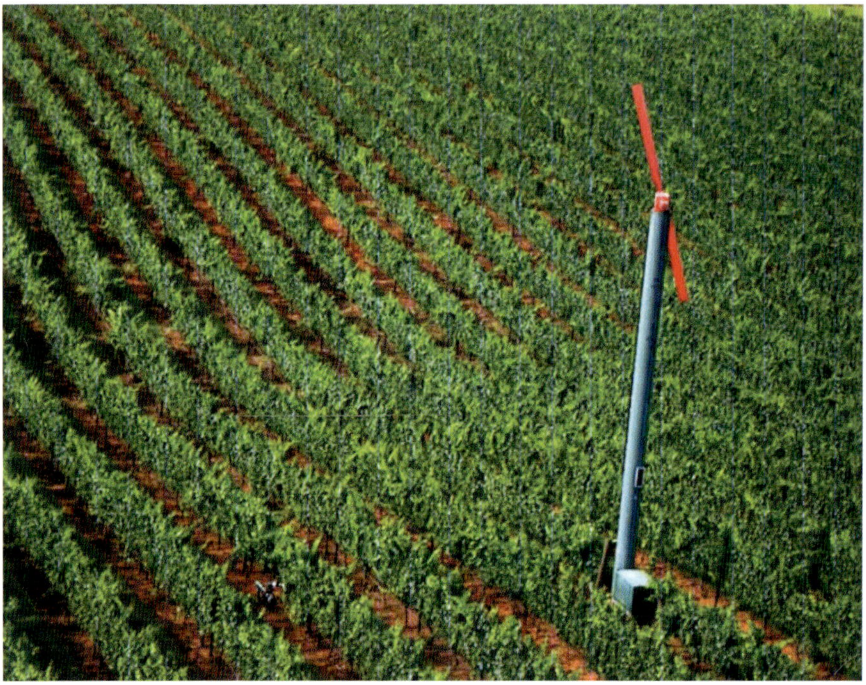

Vineyard rows with wind machine. Wind machines protect the tender grape buds in the spring when there is a likelihood of frost. Each wind machine covers ten to fifteen acres. The size of the vineyard, then, determines the number of machines needed. Shelton Vineyards in Dobson, North Carolina, for example, has thirteen wind machines. *Courtesy of George A. Denka.*

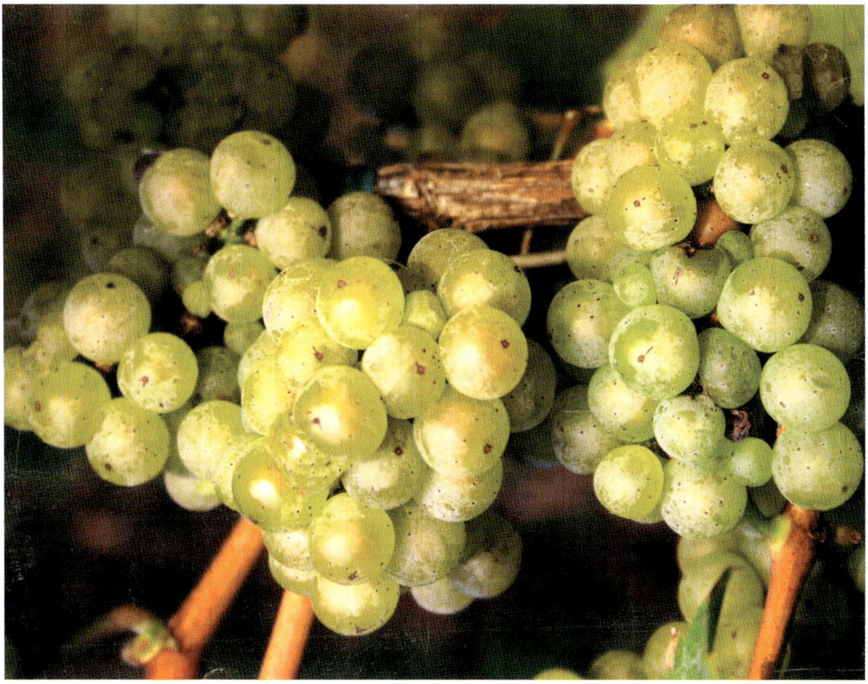

Riesling grapes. Riesling is a prized German grape but is susceptible to humidity and temperature spikes. Growing riesling grapes in North Carolina, therefore, is a challenge. Shelton Vineyards cultivates thirty-two acres, the "largest riesling planting in North Carolina." *Courtesy of George A. Denka.*

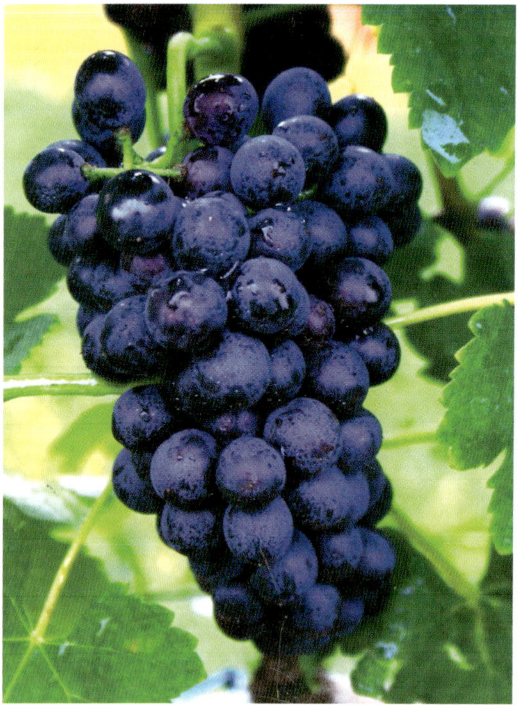

Syrah grapes. Syrah is a dark-skinned grape from France that produces a red wine. Widely cultivated, syrah is one of the top ten grown grapes in the world. *Courtesy of George A. Denka.*

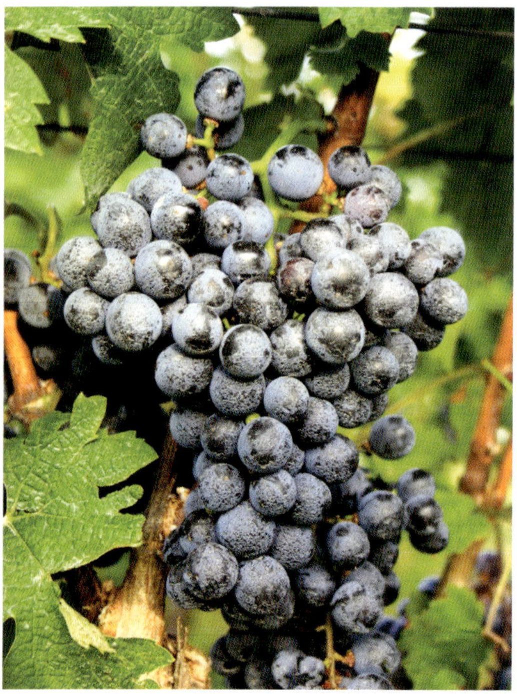

Cabernet Franc grapes. Cabernet Franc is a European cultivar (*Vitis vinifera*) grown by North Carolina estate wineries. Vintners use this grape to produce red wines. *Courtesy of George A. Denka.*

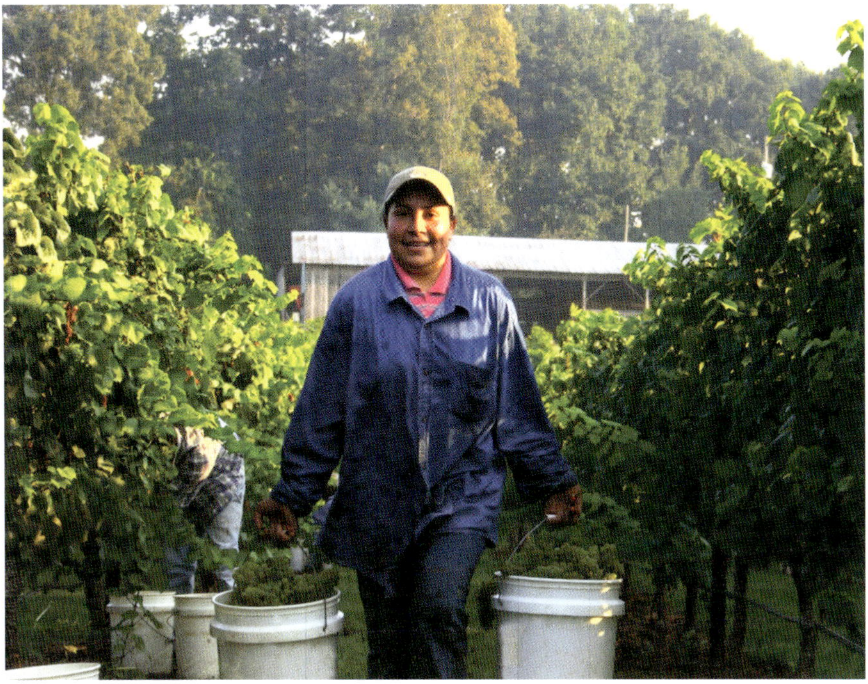

Early morning harvest, Shelton Vineyard. Harvest is a critical time for North Carolina vineyards. Many North Carolina wineries celebrate with harvest festivals in the fall. *Courtesy of George A. Denka.*

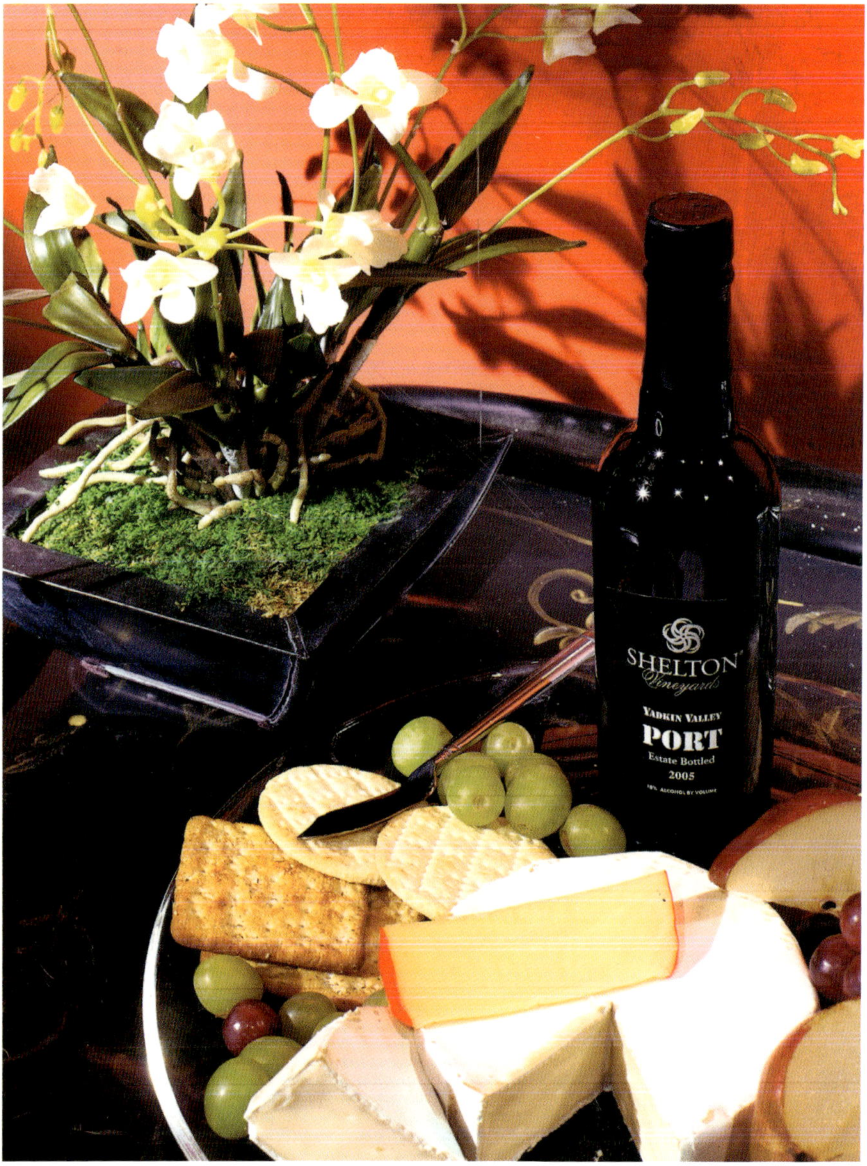

Shelton Vineyards port. Colonial and antebellum North Carolinians prized port, a fortified wine imported from Portugal. Early American wine drinkers frequently compared native wines with their favorite Old World wines. *Courtesy of George A. Denka.*

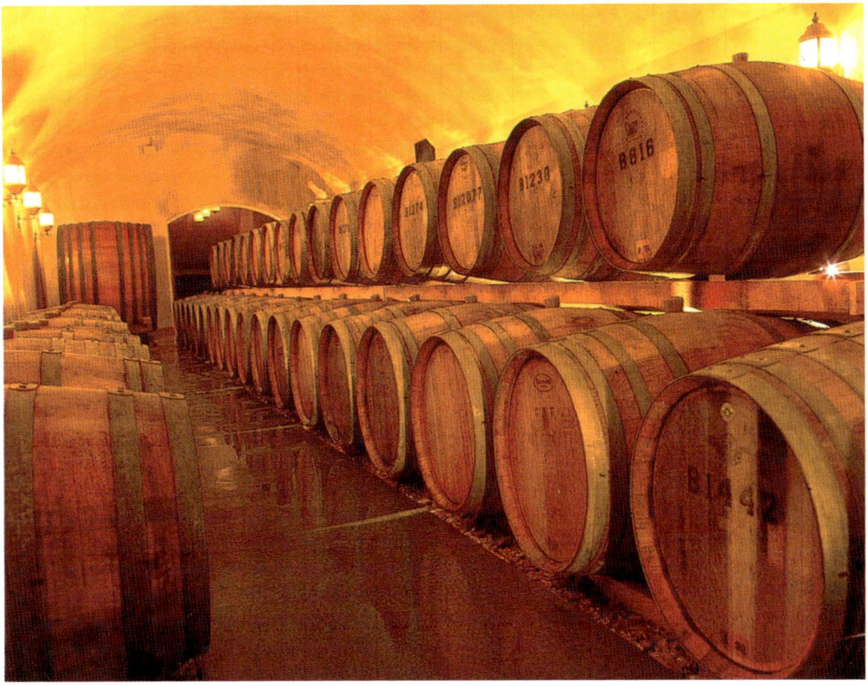

Cave with wine kegs. The word "cave," derived from the Portuguese, is a special area, often underground, with cool temperatures and high humidity for the optimum storage and aging of wine. Chardonnay is one wine that is often aged in oak barrels. *Courtesy of George A. Denka.*

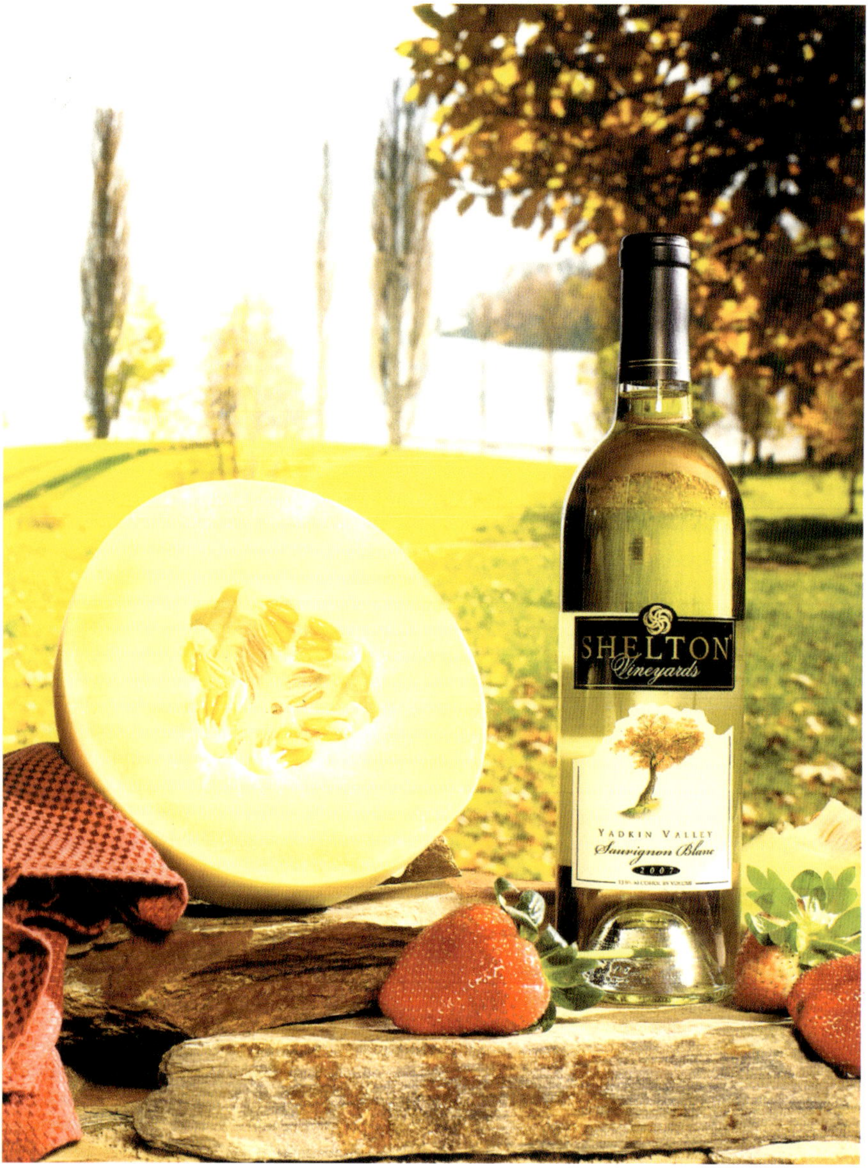

Shelton Vineyards Sauvignon Blanc. In 1999, Charles and Ed Shelton established Shelton Vineyards near Dobson. Sauvignon blanc, "wild white," is a green-skinned *vinifera* grape from Bordeaux, France. *Courtesy of George A. Denka.*

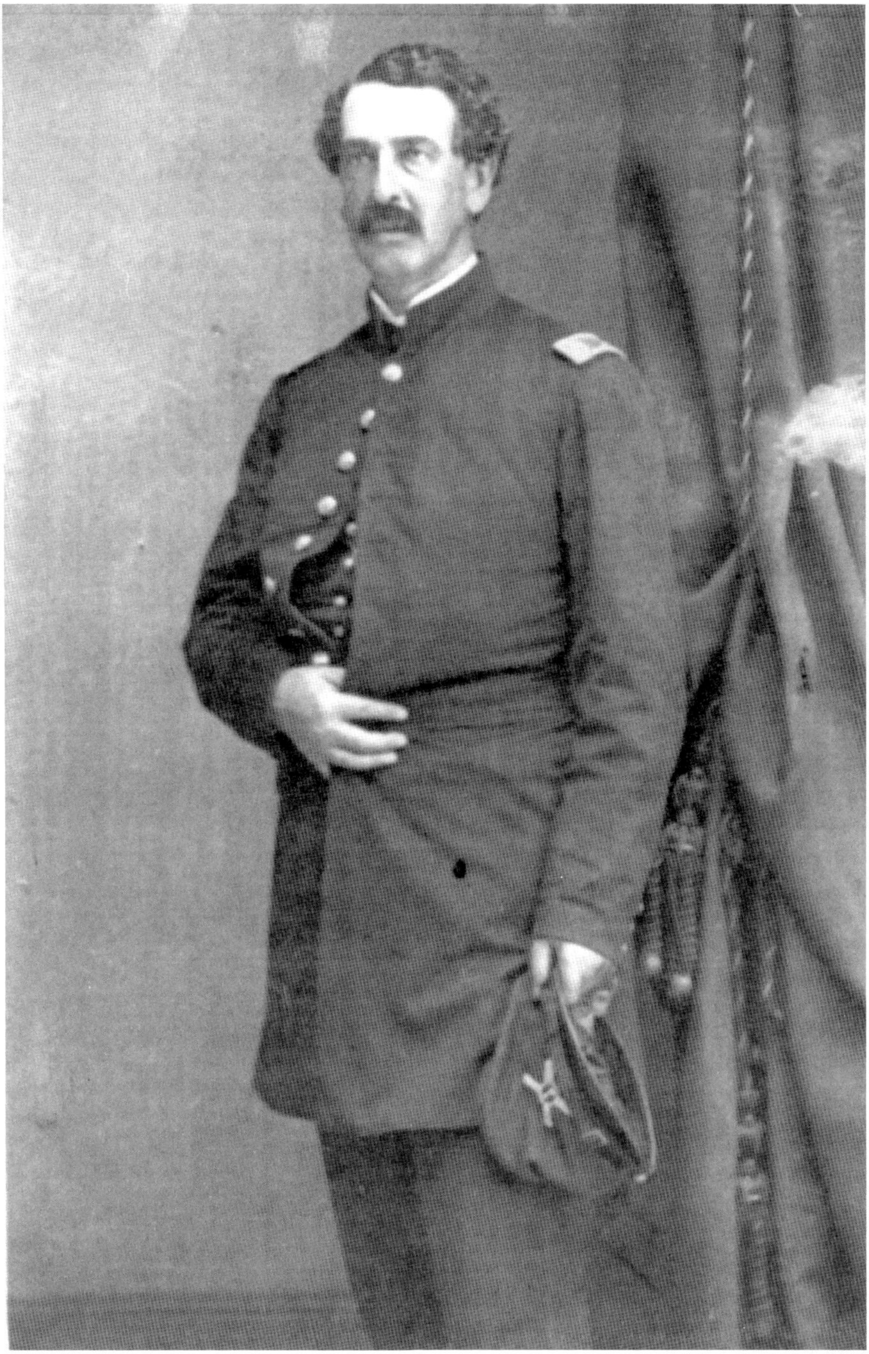

General Abner Doubleday is well known as the putative father of the great American pastime: baseball. In the 1890s, his brother General Ulysses Doubleday founded a vineyard in Polk County. *Courtesy of the Library of Congress.*

his more famous poems was "The Marshes of Glynn." Lanier died of tuberculosis near Lynn, Polk County, North Carolina, in 1881. The Doubleday vineyard featured Niagara and Delaware grapes. The 1900 census of Tryon Township listed Harold Doubleday, a native of New York, as a grape grower.

Alixes J. Lemort, a French horticulturist, helped establish the Doubleday and other area vineyards and then established his own vineyard near the blockhouse. For the 1910 census, Lemort listed his occupation as fruit farmer. By 1914, Lemort was operating the Tryon Vineyard, retailing "Pure Native Wines and Table Grapes."

A sample of the 1900 Federal Census provided insight into some of the individuals involved with Polk County vineyards. The census listed Harold Doubleday ("grape grower"), Robert H.W. Alston ("grape growing"), Thomas and John Fields (grape salesmen, perhaps to train

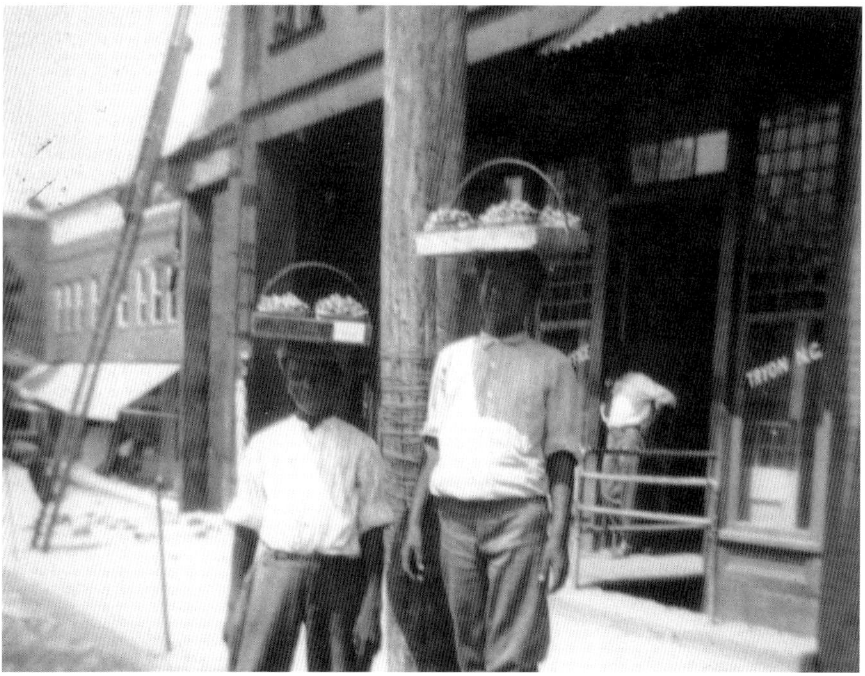

Grape peddlers, Tryon, North Carolina. Young men met the train in Tryon and other rail stops and sold grapes through the windows. This photograph was taken at the post office. *Photograph by Terry L. Helsley, courtesy of the Polk County Historical Association.*

passengers), George Morton ("horticulturalist"), Alex D. Beatson ("fruit culture") and William T. Lindsey ("fruit grower"). In 1899, Polk County ranked third in North Carolina for the number of its planted grapevines. Moore County was first, and Halifax County was second.

Many, but not all, of the 1900 grape growers in Tryon Township were northerners. An exception in 1900 was Robert H.W. Alston, scion of illustrious Georgetown, South Carolina rice planters, who also engaged in "grape growing" in Tryon Township. While a few individuals established wineries and produced wine, many chose to only plant vineyards and cultivate grapes for market.

In addition to Polk County, farmers and investors in other western North Carolina counties also explored grape growing and wine production. For example, in 1883, John K. Hoyt, a native of Beaufort County, North Carolina, bought 107 acres in Upper Hominy Township in Buncombe County, about fifteen miles west of Asheville. He then moved his family there from eastern North Carolina. Reportedly, the family lived in a log cabin while Hoyt built his mansion, Engadine. Engadine, built in 1885,

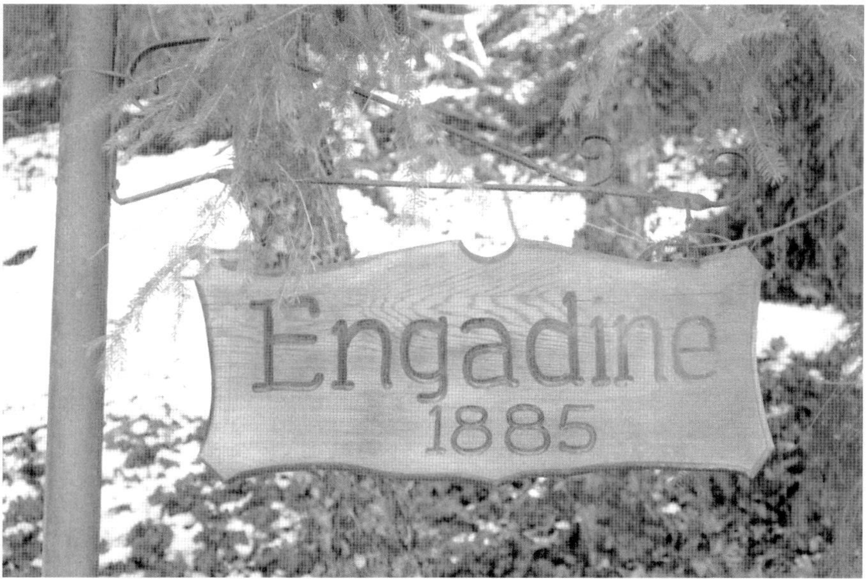

Engadine 1885 sign. John K. Hoyt built the Engadine house in 1885. The house and his winery were located west of Asheville. *Courtesy of Terry L. Helsley.*

had generator-produced electricity, running water, central heating and an earthen cellar "refrigerator" under the house. Hoyt named his home for a beautiful valley in Canton Graubunden in Switzerland. Originally, Engadine was ideally located for commerce and transportation. The winery stood near a railroad line and the Buncombe Turnpike. Today, the house, which is listed on the National Register of Historic Places, stands on Smokey Park Highway in Candler.

During the Civil War, Hoyt, born circa 1840, served in the Third Alabama Infantry, organized in Montgomery, Alabama. After the war, he moved to New York City. There he married, and in 1880, he worked as a clerk for a mercantile firm and lived in Kings (Brooklyn), New York. Eventually, the Hoyt family included John K.; his wife, Mary Elizabeth, née Bensel; her three children, Bell, Marie Louise and Minnie Brush; and their daughter, Mary B. At the time of her death in 1938, Mary Elizabeth Hoyt lived in Biltmore Forest, Asheville.

Hoyt produced a "wine of very high repute" that he exhibited with pride. Nevertheless, despite his early success, in 1899, Hoyt sold his 107 acres in Upper Hominy and, by 1910, the year of his death, had returned home and was a businessman in Beaufort, Carteret County, North Carolina.

Also near the Blue Ridge, Jules L. Golay, a native of Switzerland, initially planted his Old Fort Vineyard in McDowell County with European grapes. When that experiment failed, Golay replanted his vineyard with native grapes. The 1900 census listed Golay as a seventy-three-year-old farmer who had immigrated to America in 1879. Of interest, the census also listed a younger Jules Golay, age thirty-three years, also a native of Switzerland, as a horticulturalist in Columbus, Polk County. According to the census, the younger Golay immigrated in 1898.

In 1883, the North Carolina exhibit at the Boston Industrial Fair prominently featured a wine display as part of its agricultural exhibit. The wine exhibit showcased several varieties of North Carolina wine, including scuppernong, muscatel, Mish and claret, as well as brandy and other wine products. Wine exhibitors, according to Charles Gohdes, included Tokay, Medoc, the Seaboard Wine Company and S.R. Hunt and Company, possibly owned by Samuel R. Hunt of Kittrells in Granville County.

The 1890s were good years for North Carolina wineries. Sol Bear & Co., Tokay, Engadine, Lindsey, Doubleday and many other vineyards east and west flourished. In August 1894, a $100,000 vineyard opened in North Carolina. In 1896, North Carolina white scuppernong wine sold in Washington, D.C., for $1.50 per gallon. At its zenith in 1900, this second flowering of the wine industry in North Carolina had twenty-five wineries. North Carolina winemakers had adopted the wisdom of J.D. Cameron, who wrote in the North Carolina State Board of Agriculture's *Handbook* in 1893: "It should not be forgotten that the American grape, however improved and refined, has character and flavor of its own, and the attempt to imitate the European wines in full perfection only results, like all imitations, in failure or imperfection."

With Sidney Weller at Medoc and Wharton Green at Tokay, the muscadine grapes and their wines had come into their own. North Carolina was producing premium muscadine wines, and its wineries and grape growers prospered. By 1916, Garrett and Bear were buying three-fifths of the North Carolina muscadine crop for wine. Yet the downward slide began almost immediately.

Despite the storm clouds looming, in 1905 Stanley G. Fowler celebrated "The Scuppernong" in verse. Fowler alleged that the scuppernong was "the finest and best that gods or men ever tasted before." His last verse also celebrated the grape's health benefits, a recurring theme in scuppernong history:

From grapes so propitious, the wine, how delicious!...
New life it imparts to weakening hearts,
Gives health to the sick and joy to the well,
So wholesome and good, no laudation could
Suffice its manifold merits to tell.

Spurred by progressive ideology, many Americans embraced a political solution for social ills, including spousal abuse, poverty and lax morality. Temperance was one proposed solution that resonated with many North Carolinians. As North Carolina counties embraced temperance, Garrett left the state, and other wineries closed or consolidated. North Carolina's decision to adopt Prohibition in 1909

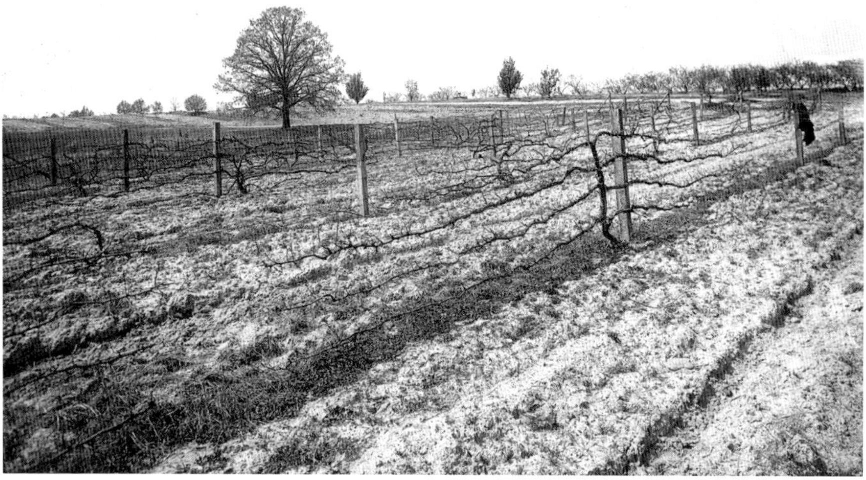

North Carolina grapevines, early twentieth century. Prior to Prohibition, North Carolina enjoyed a renaissance in vineyard plantings and wine production. *Courtesy of the North Carolina State Archives.*

sealed the doom of this second flowering of this seemingly natural North Carolina manufacture.

Consequently, the early 1900s were difficult years for the North Carolina grape and wine industries. Prohibition became the norm; vintners left the state, and eventually, with diminishing markets, North Carolina farmers looked to other crops. With the departure of Garrett and the end of Sol Bear & Co., winemaking left eastern North Carolina until the 1950s.

Chapter 4

Doom and Bloom

Hope for what we may have in future years.
—Henry Scadin

Death for the North Carolina wine business was Prohibition. Yet even before the United States ratified the Eighteenth Amendment in 1919, many North Carolina counties went "dry"—that is, outlawed the sale, consumption or production of alcoholic beverages. The State of North Carolina adopted Prohibition in 1909. In 1912, Paul Garrett, who had previously transferred his winemaking to Virginia, once again moved his operations to New York State. In the end, Garrett had seventeen winemaking facilities, plus vineyards in New York, California, Ohio, Missouri, Virginia and North Carolina.

The Eighteenth Amendment to the United States Constitution outlawed "the manufacture, sale, or transportation of intoxicating liquors within…the United States…for beverage purposes." The ban took effect on January 16, 1920, inaugurating an almost fourteen-year period in which making, selling and transporting liquors was illegal. Even before national Prohibition, half the country, including North Carolina, had already adopted the practice. While the Eighteenth Amendment stated a clear Prohibition, it was the Volstead Act of October 28, 1919, that clarified and defined the scope of the amendment. For example, the Volstead Act defined the banned substances as "beer, wine, or other intoxicating malt or vinous liquors" that were more than 0.5 percent alcohol by volume. The act also forbade the ownership of all equipment used to produce the outlawed beverages, except for non-inebriating fruit juice and cider for "home consumption." Despite these prohibitions, there were loopholes. For example, doctors could prescribe alcoholic beverages for medicinal purposes, and it was still legal to manufacture sacramental wine. Section 29 of the Volstead Act permitted the home manufacturing of "non-intoxicating" wine and cider but not of beer. This loophole provided opportunities for grape growers and manufacturers such as Garrett & Company to market juice and supplies for home manufacture. Home winemaking became an important market venue.

With passage of the Eighteenth Amendment, the foes of Prohibition began organizing for repeal. The promised end to societal ills did not materialize. Organized crime grew and flourished. Bootlegging and speakeasies grew apace. Many contended that alcohol production and consumption was a local or state matter, not a national one. Black Thursday, the Stock Market Crash of 1929, signaled the onslaught of the Great Depression. To some, repeal of the Eighteenth Amendment was a way to create jobs and improve state and federal revenue collections. Consequently, the Twenty-first Amendment to the Constitution, which repealed Prohibition, was ratified on December 5, 1933.

With the federal repeal of Prohibition, the North Carolina legislature narrowly passed an act allowing the production and sale of light natural wines. So, in time, vineyards expanded and wineries slowly returned to North Carolina. While the number of North Carolina vineyards declined under Prohibition, this important segment of the state's agriculture did not disappear. So the state was poised once again for recovery. After Prohibition ended, New Hanover was one of the first North Carolina counties to permit the sale and consumption of alcohol.

However, initially, it was not North Carolina but rather North Carolina's native son, Paul Garrett, who profited most from the repeal of Prohibition. In 1933, when the Twenty-first Amendment was ratified, only Garrett was in a position to readily produce and sell wine. During Prohibition, Garrett had diversified production and kept his plants open. For example, he manufactured Virginia Dare Tonic and Virginia Dare Flavoring Secrets (such as vanilla, cocoa, almond and many more), which in 1923 became the Brooklyn, New York–based Virginia Dare Extract Company. The Virginia Dare Extract Company is still in business. He experimented with the soft drink market and nonalcoholic wines and sold products for the home winemaker.

Prior to Prohibition, Garrett had seventeen plants in several states. So he seized the opportunity and marketed his wine in all legal states. His advertising flair rivaled that of Colonel Elliott Springs of Springs Mills in South Carolina, who was famous for his Springs Maid ads. Garrett's ads appeared in the *New York Times* as well as state and local venues. He developed singing ads and, later, television commercials. His Virginia Dare wines enjoyed phenomenal success. Garrett's post-Prohibition success hinged initially on North Carolina grapes, but the wine was produced and bottled elsewhere. In time, the scuppernong juice only flavored the wine as the demand far outstripped the scuppernong grape supply.

As early as 1905, Garrett & Company advertised their wines "as staple food for the home table." Paul Garrett even attempted to have state legislatures declare scuppernong wine and other light alcoholic beverages as "food." Such a designation would free these wines from liquor taxes and, according to Garrett, reduce unemployment. In addition to Virginia Dare, Garrett & Company also sold wine under such labels as Minnehaha and Pocahontas. Advertisements proclaimed that Garrett's scuppernong wine freed the consumer from the "bondage of tea and coffee" as it supplied "an adequate food supply for starved nerves." Other Garrett ads had a patriotic flavor, as they contended that drinking Garrett wines freed "America from dependence upon foreign countries for its wines." Paul Garrett, the great champion of American grapes and their wines, died in 1940. His successors could not build on his achievements, and in time, his heirs liquidated Garrett's vast holdings.

Established circa 1930, the Old Mother Vineyard on Roanoke Island was one of Paul Garrett's suppliers. Later, Mother Vineyard was a brand name for wine originally produced from scuppernongs grown there. In 1956, the trademark changed hands, and the label was no longer site-specific.

Polk County in western North Carolina also experienced a post-Prohibition revival. Garland Goodwin of Polk County reported that

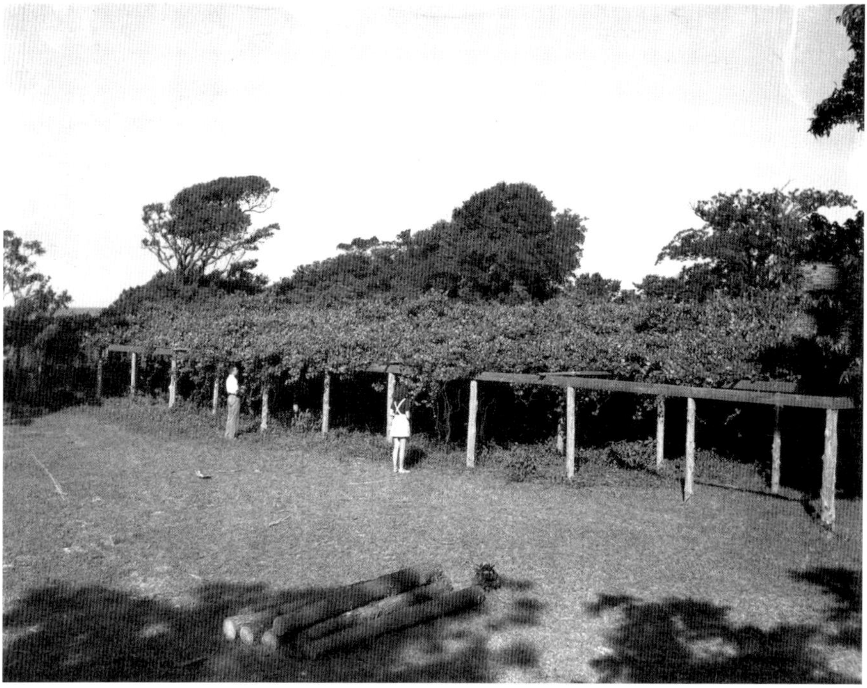

Mother Vineyard, Manteo, North Carolina. *Courtesy of the North Carolina State Archives.*

his father, also named Garland Goodwin, had landscaped and planted vineyards at Stone Hedge estate and for E.G. Vollmer on Warrior Mountain. Goodwin remembered when there were three huge green areas—the sites of the former vineyards on Warrior Mountain. The younger Goodwin entered the printing business and printed two-color labels for bottles of Vollmer wine. According to Goodwin, they used gummed paper for the labels. They cut the paper and then printed the labels in one color. As the gummed paper curled up after the first run, the printers had to wait for it to uncurl before they could print the second color. According to Goodwin, Polk County was "dry," and many, including his mother, "thought they shouldn't be making wine from the grapes." Today, the Warrior Mountain vineyards are overgrown.

Ernst G. (1884–1957) and Emma Louise Rebmann Vollmer (1893–1963) immigrated to the United States from Gerlingen, Germany. The family arrived in Tryon on Thanksgiving Day 1924. Dr. Benno von

Kahlden, a private school instructor, sponsored the Vollmers. Kahlden needed help with his vineyard near Tryon. In 1930, the Vollmers acquired six acres on Vineyard Road near Tryon, Columbus County. There, the Vollmers planted grapes and trees and established Vollmer's Vineyard. Their son, John Rebmann Vollmer, wrote that "each grape vine was like their own child and they cared for them tenderly." In time, the Vollmers established their own winery and produced popular wines such as Tryon Rhinegold Wine and Light Tryon Red Wine. Sadly, Ernst Vollmer suffered a heart attack in 1933 and, although he lived many more years, was unable to work his beloved vineyard.

Another Polk County resident, Louis Fuldner, owned the Falerno Vineyard on Meadowlane Drive on Warrior Mountain and bottled, among other proprietary wines, Tryon Sauterne. In 1930, Dr. Fuldner, a native of Minnesota, was fifty-seven years old. He lived in Tryon Township with his wife and brother-in-law, both natives of Austria. In 1939, Fuldner advertised in the *Tryon Daily Bulletin* "Pure Natural Wines" from the Falerno Vineyards Winery. Also, reputedly, Dr. Juanita Lee of Polk County had a license to make sacramental wine in the 1930s. So she and Fuldner may have exploited exceptions in the Prohibition legislation to produce small quantities of wine for medicinal or other exempt markets. During Prohibition, many patients asked doctors for

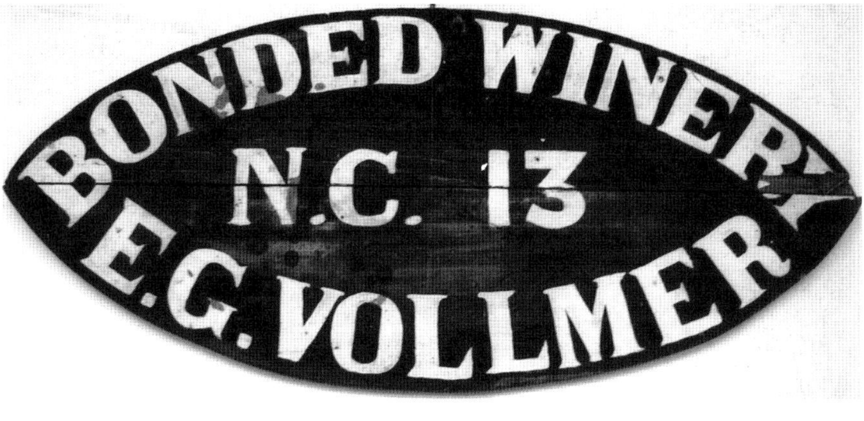

E.G. Vollmer Winery sign, Polk County, North Carolina. Ernst Vollmer, a German immigrant, operated a vineyard and winery on Warrior Mountain. *Photograph by Terry L. Helsley, courtesy of the Polk County Historical Association.*

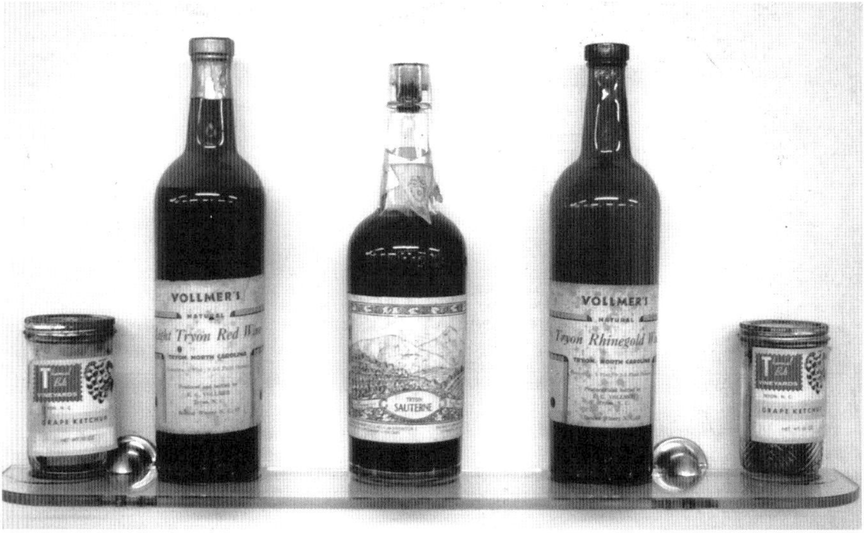

E.G. Vollmer Winery and other Polk County, North Carolina wines. Polk County in western North Carolina was a center for grape growing and wine production in the late nineteenth and early twentieth centuries. *Photograph by Terry L. Helsley, courtesy of the Polk County Historical Association.*

wine "prescriptions." Many considered wine an aid to digestion. From ancient times, man has appreciated the therapeutic qualities of wine.

Grape growing was a popular enterprise in Polk County. In the 1920s, Tryon grapes were "famous the world over." The site of Tau Rock Vineyard, one of the many that operated in the area, was off Highway 176. In 1937, Tau Rock advertised Tryon grapes: "Tryon-True, red, white and blue, 'the finest grape that ever grew.'" The accompanying story indicated that a late winter and dry, hot summer had produced a short grape season. Customers were urged to act quickly and advised that they could pick up or order grapes during the evening in the packing shed on the Tryon-Saluda Highway. Other grape growers also sold grapes from stands along the highway, and according to Polk County native Dr. George A. Jones, the vendors did a "huge business." By the 1950s, Tryon grapes were still prized for their "delicacy and flavor," but most were produced for a generally local market.

Mother Vineyard Winery was organized about 1930 and operated in Manteo until 1954. It used scuppernong grapes from the "Mother Vine" on Roanoke as well as muscadines from other vineyards for its

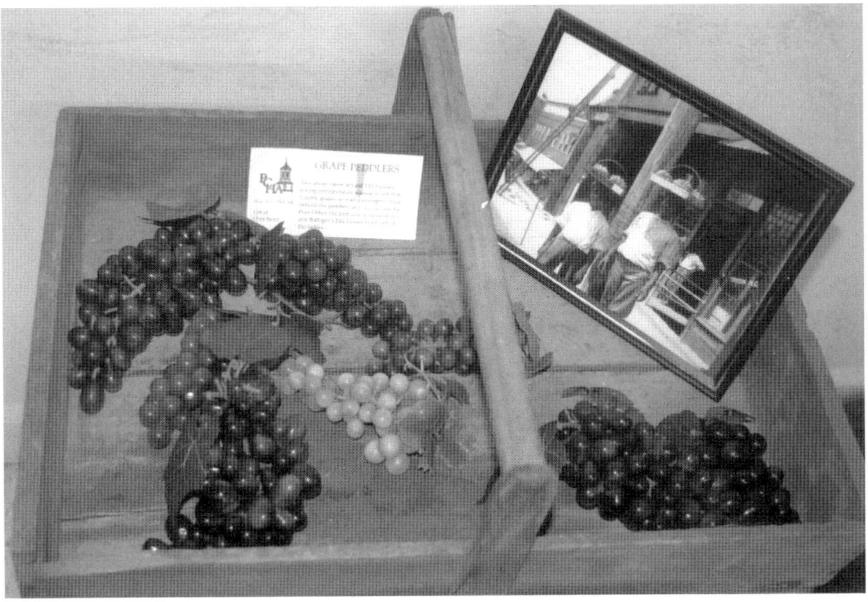

Grape peddlers' basket, Tryon. According to a Polk County native, peddlers who met the trains often sold more than produce. In addition to those who sold their wares through the train windows, some enterprising young men, known as "butch boys," also jumped on the trains peddling produce, crafts and sandwiches. *Photograph by Terry L. Helsley, courtesy of the Polk County Historical Association.*

wines. Canandaigua purchased the Mother Vineyard Winery in 1948. The Old Mother Vineyard also supplied grapes to Paul Garrett. Under Garrett, Mother Vineyard became a bestselling brand of scuppernong wine. A company in Petersburg, Virginia (a subsidiary of Canandaigua), also acquired the rights to the "Mother Vineyard" name in 1956. Former state senator Fountain Odom in 2003 established Mother Vine LLC to cultivate clippings from the Roanoke vine and produce wine. The resulting Mother Vine Premium Scuppernong Wine was, according to David Fussell Sr., "one of the most delicious wines you have ever tasted."

In 1935, Tenner Brothers Winery opened in Charlotte, Mecklenburg County. The Tenner brothers—Sol, Albert and Leon Lukie—were natives of New York City. Charleston businessmen for a time, the Tenner brothers relocated to Charlotte, opened a restaurant (with a sister) and developed a winery. They produced muscadine wines, but by 1958, the winery had moved across the state line to Patrick, South Carolina. Tenner Brothers Winery bottled its wine under the popular "Old Mule" label

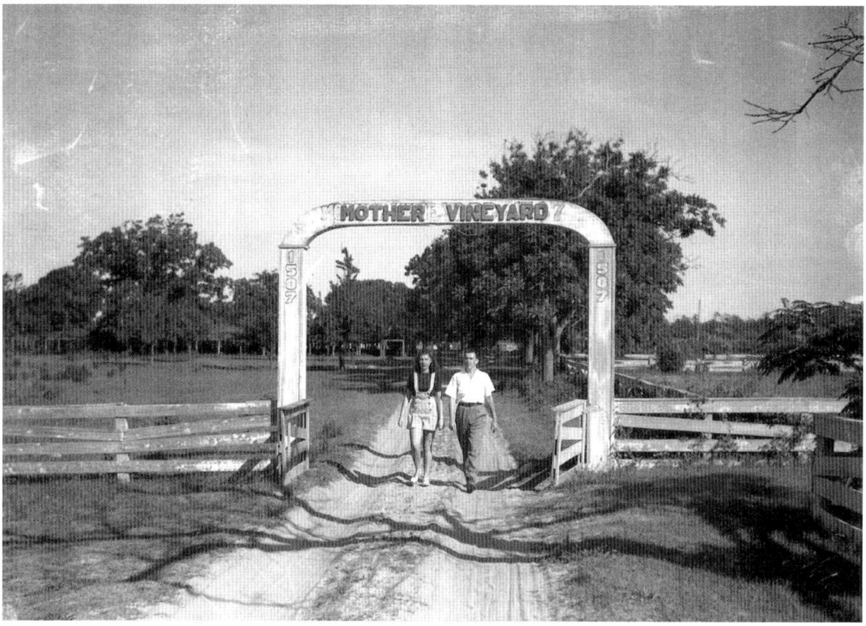

Mother Vineyard entrance, Manteo, North Carolina. Beginning in the 1930s, the Mother Vineyard cultivated scuppernong grapes. *Courtesy of the North Carolina State Archives.*

with the slogan "Old Mule's Got It." In 1965, Canandaigua acquired Tenner Brothers Winery.

Also in 1935, Mordecai "Mack" Sands opened Car-Cal Winery in Greensboro, Guilford County. The winery produced native wines for local distribution. Car-Cal wines bore, among other labels, the Old Maude label. Ads proclaimed, "Here comes Maude." Sands imported grapes from California in railcars. Car-Cal closed in 1948 when Guilford County went dry.

Marvin Sands, the son of Mack Sands, studied the wine business and, in 1945, established Canandaigua Industries in a converted sauerkraut plant in Canandaigua, New York. In 1966, Canandaigua acquired the rights to the "Virginia Dare" name. The company's most profitable early label was Richard's Irish Rose. In addition, among other labels and wineries, Canandaigua wines also included the Mother Vineyard label. Canandaigua Industries, now Constellation Brands, with hundreds of labels and wines, is one of the largest wine conglomerates. Richard Sands, the third generation, is chief executive officer of Constellation Brands.

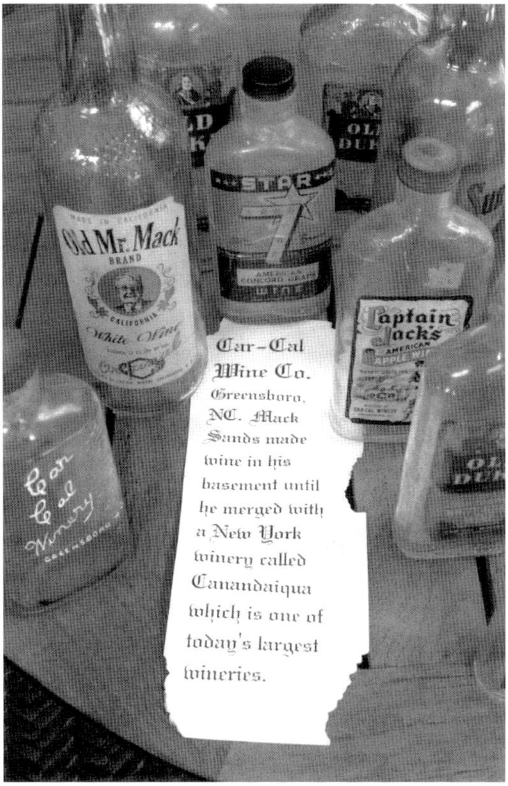

Car-Cal Wine Co., Greensboro, North Carolina bottles. In 1936, Mack Sands opened Car-Cal Winery with a limited distribution. The name Car-Cal represented Carolina-California. The winery closed in 1948. *Photograph by Terry L. Helsley, courtesy of David G. Fussell Sr.*

Mellie Bernard started his winery in 1935 in Icard Township, Burke County. Bernard was a native of Alabama and, in 1930, was the proprietor of a filling station. The winery had a storage capacity of 130,000 gallons but closed in 1948 when Burke County went dry. Bernard's wines carried the Bernard's Waldensian label. He produced a well-received scuppernong wine.

Despite a post-Prohibition business upturn, life in the 1930s was difficult for North Carolina farmers. The South had been in an agricultural depression since the advent of the boll weevil, an insect that infected cotton plants. Without cotton, small farmers sunk deeper into debt and poverty. To address this problem, Charles Garrett proposed government support to convert poor cotton farmers into grape growers. He lobbied the Rural Rehabilitation Agency to support native grape cultivation as an option for poor southern farmers. Several North Carolina counties, including Bladen, Columbus, Duplin and Samson, participated in the program.

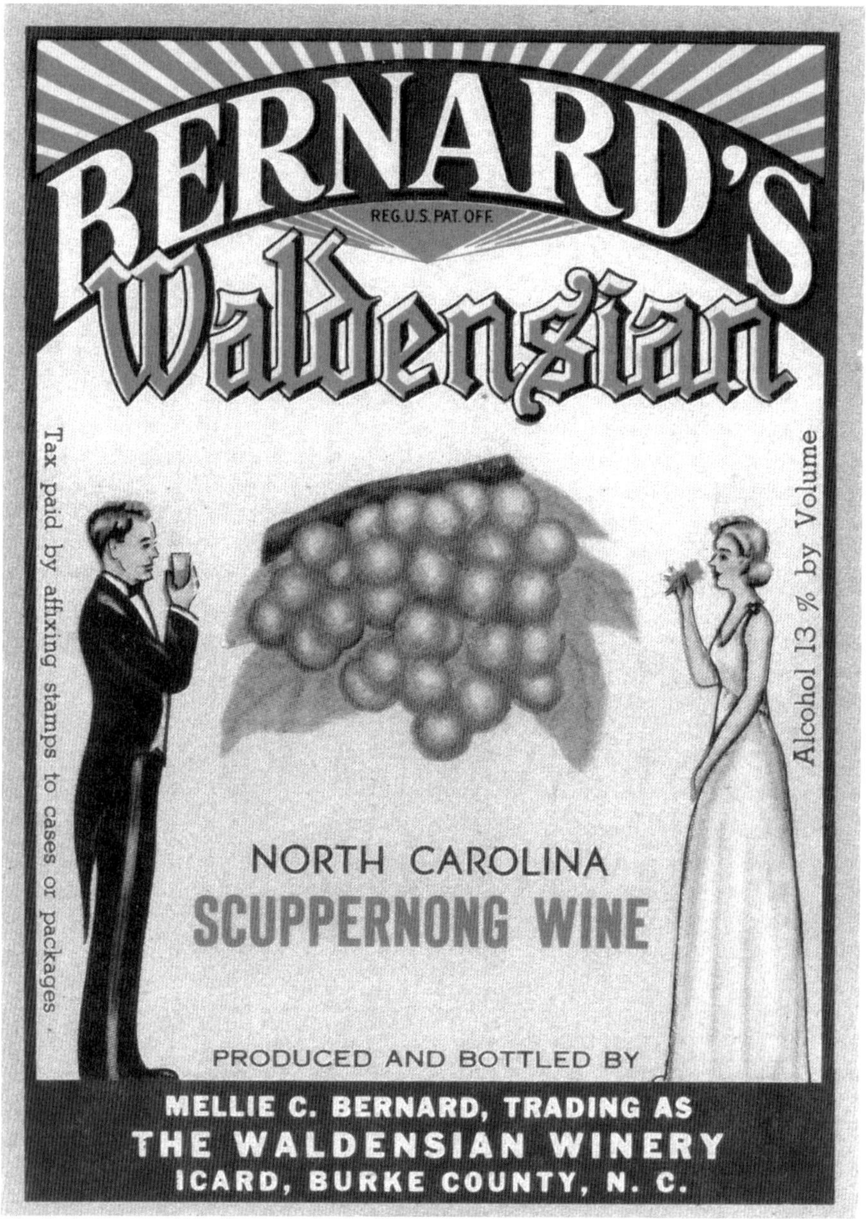

Bernard's Waldensian, North Carolina scuppernong wine label, Burke County, North Carolina. Mellie Bernard started his winery in 1935 in Icard. The winery had a storage capacity of 130,000 gallons but closed in 1948 when Burke County went dry. The original label showed a man and a woman holding wine glasses. It was rejected because it was "not lady-like for women to have a glass of wine in public." So the woman held a rose instead. The Waldensian name was possibly a tribute to a group of Waldenses, Reformed Protestants, from Italy who settled in eastern Burke County near the Catawba River in 1893. *Courtesy of David G. Fussell Sr.*

Yet despite a few fledgling wineries, North Carolina was more a source of raw materials—in this case grapes—than a manufacturing center. Grape growing in early twentieth-century North Carolina was basically in the east and in thermal pockets in western North Carolina. It was an important option but not a rapidly increasing one for the state's farmers. North Carolina farmers grew primarily muscadine grapes for wineries in other states, and in 1930 there were only 383 mature (that is capable of bearing fruit) grapevines in North Carolina. The pendulum, however, was beginning to shift. For example, in 1950, farmers in Onslow County planted twenty-five acres of scuppernong for an out-of-state market. This effort floundered in 1955 when the out-of-state winery refused to buy the grapes. Reacting to this situation, Raymond A. Hartsfield, one of the growers, decided to build a winery—Onslow Wine Cellars—to provide a local market for the grapes. Hartsfield's father had also been in the wine business.

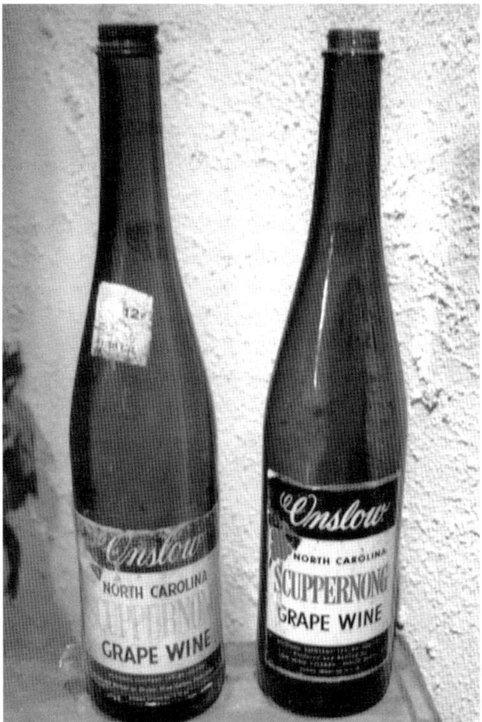

Onslow scuppernong wine bottles. In the 1950s, Raymond Hartsfield established Onslow Wine Cellars in order to provide a market for Onslow County grape growers. *Photograph by Terry L. Helsley, courtesy of David G. Fussell Sr.*

Hartsfield, the son of N. Allan and Louisa Hartsfield, expanded his operations as demand for scuppernong wine increased. In 1961, Onslow was the only company in North Carolina commercially producing wine. In 1969, according to Thomas Pinney, Canandaigua of New York acquired Onslow Wine Cellars and, through such acquisitions, "became the largest wine producer outside California." In 2000, Canandaigua Industries became Constellation Brands, Inc.

Chapter 5

Scuppernong and Syrah

We do not see why good wines should not be made in this State.
—Wilmington, North Carolina Herald, *1851*

In 1961, a Virginia firm, Richard's Wine Cellars, a subsidiary of Canandaigua, contracted with North Carolina farmers to plant muscadine grapes. Richard's Wine Cellars furnished the vines and promised $200 per ton of grapes produced. With this incentive, many farmers turned from tobacco to grapes. The State of North Carolina was also interested in diversifying the state's agricultural production, providing agricultural alternatives to cotton and tobacco. This interest led State Senator Carl Vitners of Onslow County to introduce legislation for a state appropriation to support wine research and educational opportunities for grape growers. As a result, two professors at North Carolina State University, Dr. Bill Nesbitt and Dr. Dan Carroll, worked to improve muscadine strains and to enhance the processing of muscadine wine.

Despite the interest in and support of grape growing, after the sale of Onslow Wine Cellars, North Carolina did not have a resident winery. The state's grapes were sold and processed out of state. This situation was an economic setback for the state, as manufacturing is more profitable than supplying raw materials.

In 1972, however, the situation changed. The North Carolina legislature developed a stimulus plan to encourage the development of North Carolina

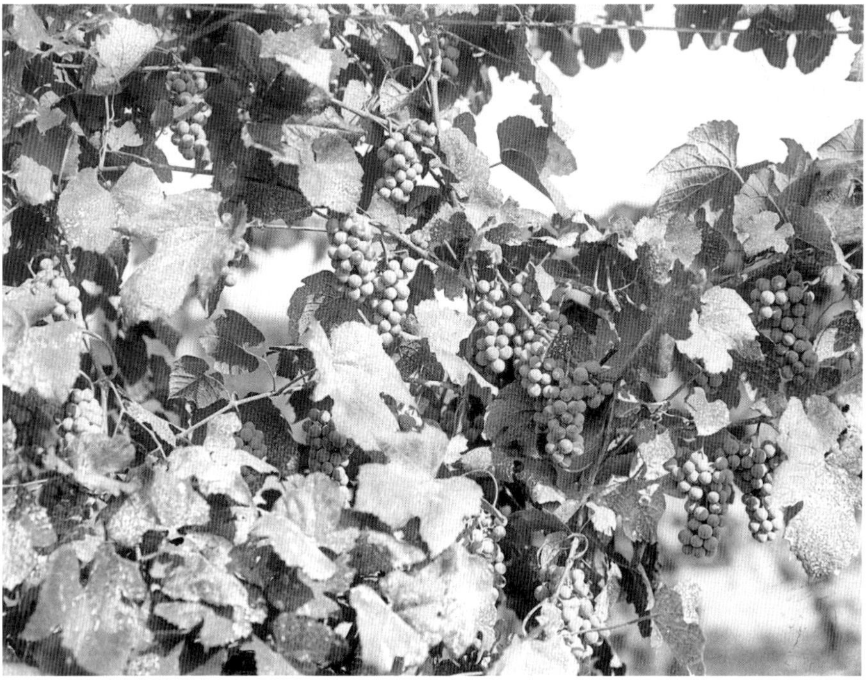

Grapes, Southern Pines, North Carolina, circa 1900. In 1893, there was a vineyard that covered several hundred acres at Southern Pines in Moore County. Moore County in the 1890s was a center of grape cultivation. Growers generally favored Niagara, Delaware, Concord and scuppernong grapes. *Courtesy of the North Carolina State Archives.*

wineries and to enhance grape growing as an agricultural option for North Carolina farmers. The legislature drastically reduced the annual license fee for a winery and cut the state tax on wines produced from native grapes.

Specifically, the State of North Carolina enacted legislation giving North Carolina wineries preferential tax status. Under this legislation, wines made from 51 percent or more native fruits (including grapes) had a state excise tax of five cents per gallon, while other wines paid sixty cents per gallon. Consequently, the New River Grape Growers Association refocused its efforts and expanded its mission as the North Carolina Grape Growers Association.

In 1972, muscadine growers organized the North Carolina Muscadine Grape Association, Inc. (NCMGA). The association promoted research, merchandising and production efficiency. The association represented individuals involved in a wide spectrum of muscadine-related enterprises.

The 1970s, then, were a crucial decade for the wine industry in North Carolina. The preferential legislation immediately impacted the North Carolina wine industry. Wineries, including Duplin Wine Cellars in Rose Hill (1976), Biltmore Estate Winery in Asheville (1978) and Germanton Vineyard & Winery in Greensboro (1981), opened across the state.

Not only did the tax incentive trigger new wineries, but it also expanded the scope of North Carolina winemaking. Visionaries at Biltmore (William A.V. Cecil) and Westbend Vineyards (Jack Kroustalis) dreamed of European-style wines in the California tradition and began planting *vinifera* vineyards. In eastern North Carolina, grape farmers embraced muscadine wine production, with Duplin Wine Cellars as a guaranteed market for their grapes. At last, the historical dualism of North Carolina wine production yielded vibrant streams of viticulture that finally propelled the Old North State into one of the great wine-producing areas in America.

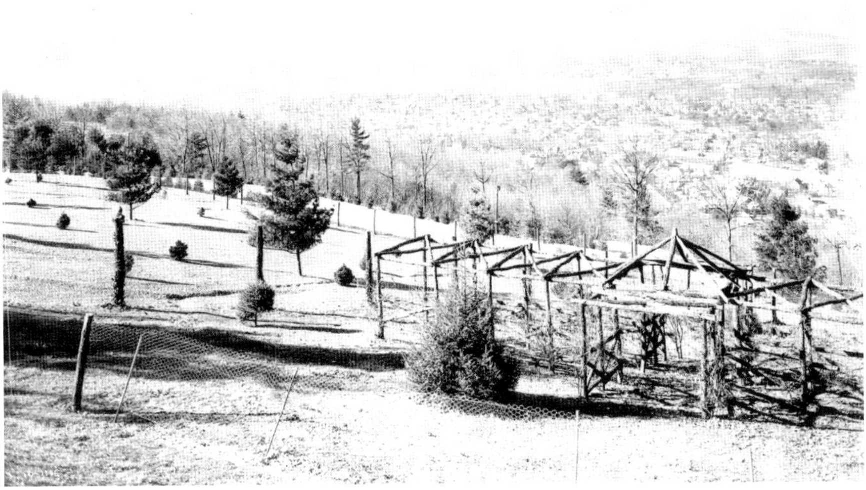

Vineyard at Biltmore with Asheville skyline. Even before the first modern vineyards were planted in 1971, the Biltmore estate had a long agricultural tradition and a commitment to sustainability. *Courtesy of the D.H. Ramsey Library, University of North Carolina–Asheville.*

Countering the long-held view that European wines were not suited for North Carolina, in 1972 Jack Kroustalis established Westbend Vineyards in Lewisville along the Yadkin River in Forsyth County. Kroustalis, despite opposition, followed the California model. He planted nine varieties of California *vinifera* and produced California-style wine. Westbend Vineyards produced the first *Vitis vinifera* wine in North Carolina. In 1986, Kroustalis and his wife, Lillian, had a seventy-ton grape harvest. Sadly, this *vinifera* pioneer died in Lewisville in 2006.

A number of growers planted scuppernong grapes in the 1970s. Some of these were Frank Hollowell of Pasquotank County, the Wood family in Chowan County, the Sell family in Brunswick County and the Fussell family of Duplin County. Between 1974 and 1980, Frank Wood and Paul Williams operated Deerfield Wine Cellars near Edenton, Chowan County. At its height, Deerfield produced fifty-five thousand gallons of wine from eighty acres of muscadines. However, in 1980, when Frank Williams, the principal owner, died, Deerfield closed. The Deerfield vintners mentored David Fussell Sr. of Duplin Winery.

The next chapter in the history of North Carolina wine focuses on the first three bonded wineries in North Carolina: Germanton (the first) near Greensboro, Duplin Winery (the second) in Rose Hill and Biltmore Estate Winery (the third) near Asheville. The most successful of these pioneers were Duplin Winery and Biltmore Estate Winery. Due to federal regulations, wineries must post a surety bond to guarantee their payment of federal excise taxes on the wine they produce. The tax is due when the wine is sold.

Members of the Piedmont Grape Growers Association organized Germanton Vineyard & Winery in Germanton in 1981. According to wine historian Leon Adams, the five vineyard owner-founders of Germanton were William McGee, Jerry Pegram, Scott Lawrence, John Gillispie and Dr. Mallory Chambliss. The winery located in a former barn seven miles north of Winston-Salem produced its wine from French and American hybrid grapes. Germanton was North Carolina's first bonded winery after Prohibition.

Duplin Winery

Duplin Winery in Rose Hill, North Carolina, stands in the tradition of Onslow and Deerfield. David Fussell Sr. and his brother Daniel Jerome Fussell Jr. made their first muscadine wine in 1975 and sold their first wine in 1976. Their parents were family winemakers. David Fussell was an educator and farmer when he entered the wine business. As David Fussell commented in a 2008 interview, to make wine "all you do is put the grapes in and stand back and watch and He [God] does it." As a winemaking novice, David Fussell adapted wine recipes, especially for a sparkling scuppernong wine, from Sidney Weller's *Southern Wine Making and Viticulture* (1836). Fussell studied and experimented for several years to develop successful wines. Dr. Dan Carroll of North Carolina State University also assisted the Fussells in developing their winery.

The Fussells and other muscadine grape producers formed Duplin Wine Cellars as an outlet for the grapes grown by eastern North Carolina farmers. Grape production had accelerated in the 1960s after Richard's Wine Cellars, a subsidiary of Canandaigua Industries of New York, offered five-year good contracts at a guaranteed return ($200 per ton) to farmers who would grow muscadine grapes for them. At that time, Richard's produced scuppernong wine under the Mother Vineyard label. As a result, the cultivation of grapes increased from about four hundred acres in 1966 to several thousand acres in the 1970s. In time, supply exceeded demand, and the price dropped to $125 per ton—too low a price for the farmers to break even. As the farmers cast about for marketing options, Duplin Wine Cellars, later Duplin Winery, was born. The Fussells and seven other grape-growing families banded together to produce a muscadine wine, according to David Fussell Jr., "that tastes just like you're up underneath the grapevine eating the grapes." However, not everyone in the community supported the move. Some referred to the winery as the "Factory of Liquid Sin." In 1979, David Fussell bought his brother's interest in Duplin Winery.

The year 1991 was a turning point for the winery. A story on *60 Minutes* reported on the so-called "French Paradox," which said that the French, who eat more fat and smoke more than Americans, had only one-third the number of heart attacks and less than half the rate of heart disease. Researchers

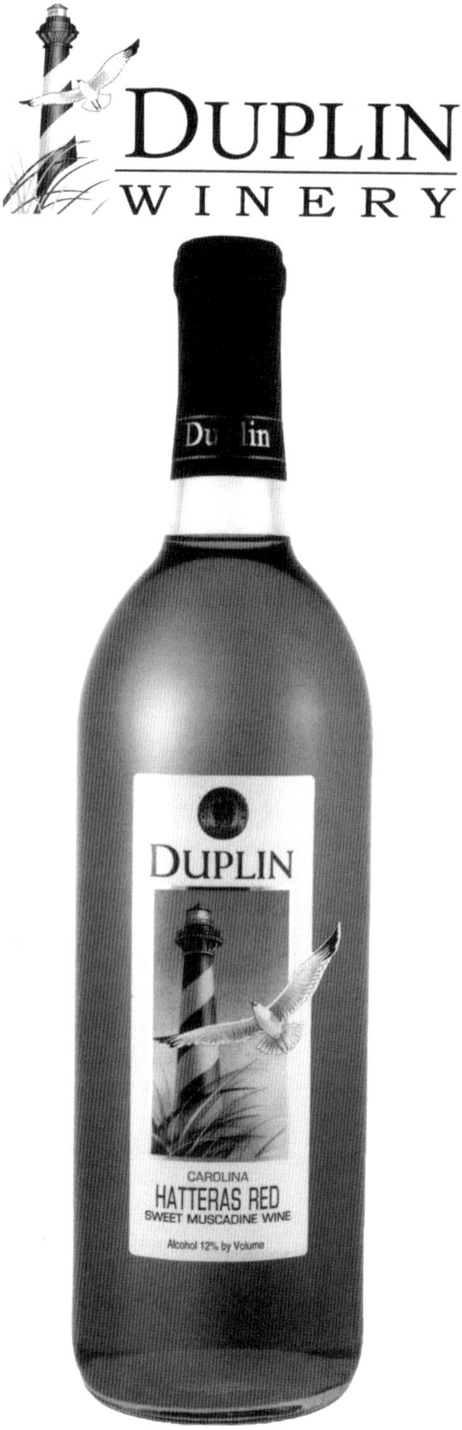

A bottle of Duplin Hatteras Red. Duplin Winery, Rose Hill, North Carolina, according to founder David G. Fussell Sr., markets "the best selling wine east of I-95 in North Carolina." *Courtesy of David G. Fussell Sr.*

suggested that drinking red wine had health benefits. Red wine had high levels of resveratrol, an antioxidant that promoted heart health. The *60 Minutes* story featured Dr. Morton Gronbaek, author of the Copenhagen City Heart Study. Gronbaek published the findings in the *British Medical Journal.* Underwritten by the Danish National Board of Health, the Copenhagen City Heart Study demonstrated the benefits of moderate wine consumption. The study followed over thirteen thousand individuals between 1976 and 1988 and discovered that those who drank wine daily were less likely to die than those who abstained or drank other alcoholic beverages.

As a result, Duplin began to produce red as well as white wines. The pharmaceutical college of Campbell University, Buies Creek, North Carolina, tested *Vitis rotundifolia* grapes for resveratrol. Scientists there found that wines made from muscadine grapes had seven times more resveratrol than other red wines. Sales improved, and in 2008, Duplin had fifty-five employees and two thousand acres of grapes and produced about 280,000 cases of wine. The concern also involved forty-three farmers who raised muscadine grapes for the operation. Duplin used such North Carolina muscadines, such as Carlos, Magnolia, Noble and Regale, to produce its wines.

In 2009, the winery produced 1.3 million gallons. Duplin Winery, a family business, is the oldest operating winery in North Carolina and the largest one in the South. Today, David G. Fussell Jr. is president of the winery, and his brothers, Patrick G. and Jonathan D. Fussell, are also involved in winery administration. So the tradition of North Carolina winemaking continues in the third generation of the Fussell family. In 2010, for the second year in a row, the Beverage Information Group presented Duplin Winery with its Fast Track Brand Growth Award. At Duplin, David G. Fussell Sr. summed up the secrets to producing good wine: "having a good taster," "having good fruit" and "having good will."

Biltmore Estate Winery

Also in the early 1970s, William A.V. Cecil, George Vanderbilt's grandson, began planting grapes on the Biltmore Estate. In 1971, he

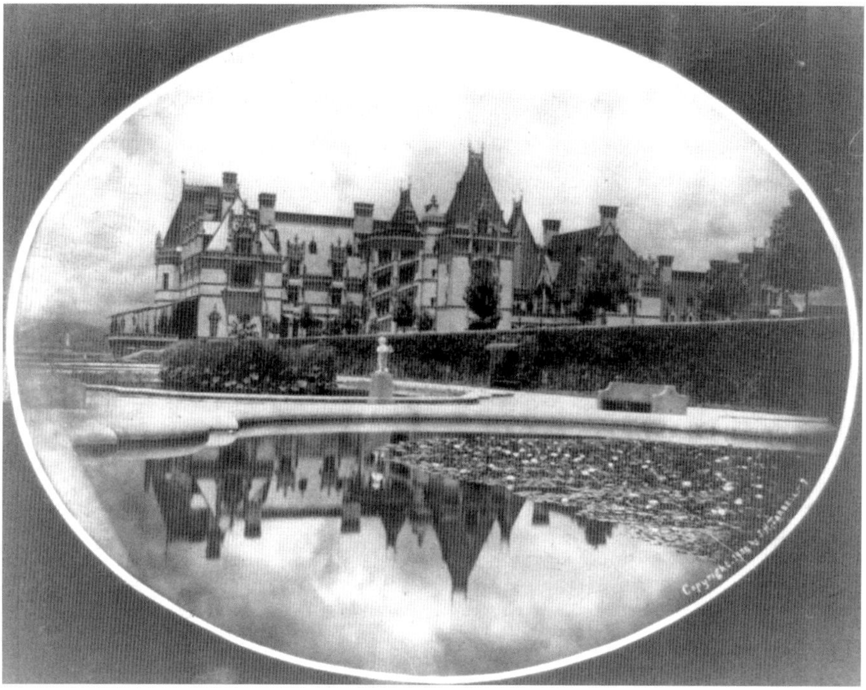

Biltmore House, circa 1900. Built for George Washington Vanderbilt in 1895, Biltmore is the largest private dwelling in the United States. *Courtesy of the Library of Congress.*

planted an experimental vineyard and in 1979 sold the first wines with the Biltmore label. The Biltmore Estate Winery opened in May 1985 in a renovated dairy barn. Several factors influenced William A.V. Cecil's decision to invest in the vineyard and open Biltmore Winery. According to Jerry Douglas, senior vice-president, these factors included the Biltmore Estate's strong agricultural heritage and commitment to sustainability and the opportunity to diversify the estate's tourism business. In addition, as the French chateaux of the Loire Valley were the models for the Biltmore house, there was a certain feeling that adding a vineyard was a natural progression. As Douglas commented, "What makes more sense than for a chateau to have a vineyard?" The Biltmore Estate overlooks the French Broad River.

At first, Biltmore planted native grapes and then moved to French-American hybrids. Eventually, the winery turned to *vinifera* grapes. *Vinifera* grapes seemed more appropriate to a French chateau, and visitors wanted chardonnay and other wines produced from the French grapes. In 1977, Cecil met and employed Philippe Jourdain, an experienced French winemaker who had studied enology, the science of winemaking. Employing Jourdain was a vital step in the development of the multimillion-dollar winery. Jourdain had the expertise to develop the quality of wine Cecil wanted for Biltmore.

Today, Biltmore Estate Winery is the "most visited winery" in the United States and controls 30 percent of the wine market in the states where its wines are available. Biltmore Estate Winery is also a major market for area grape growers. In 1994, the Biltmore vineyard of *vinifera* grapes covered ninety-four acres—according to Rachel Carley, "one of the largest plantings of *vinifera* grapes east of the Mississippi." Biltmore's *vinifera* cultivars include cabernet sauvignon, cabernet franc, merlot, chardonnay, riesling and viognier. In the 1980s, the winery also began to produce and market champagne. In 2004, the winery sold 1.4 million bottles of Biltmore wine.

Biltmore Estate Winery, according to Jerry Douglas, has had a direct and indirect influence on the North Carolina wine industry. Directly, Biltmore was the third bonded winery in North Carolina. The winery was also a pioneer in East Coast *vinifera* plantings and winemaking. Its success proved that *vinifera* grapes had economic possibilities in North Carolina and that North Carolina vineyards could produce award-winning *vinifera* wines. Biltmore actively supported efforts of the North Carolina Grape Council to promote viticulture in North Carolina and to raise the national profile for the state's wines. In addition, Biltmore Estate Winery, with its high volume of visitors, plays a major role in exposing the visiting public to the possibilities of wine—its varieties and its cost. As Douglas indicated, North Carolina wineries produce such a range of red, white, sweet, dry, sparkling and blush wines that there is a wine that fits every consumer's taste.

With the success of North Carolina vineyards and wineries, the farmers of North Carolina had another economic model. They were no longer tied to tobacco or cotton cultivation. The Biltmore success also encouraged others to pursue the planting of vineyards and winemaking in North Carolina either as a lifestyle choice or commercially. Grape growing also helps preserve farmland and the agricultural way of life in North Carolina.

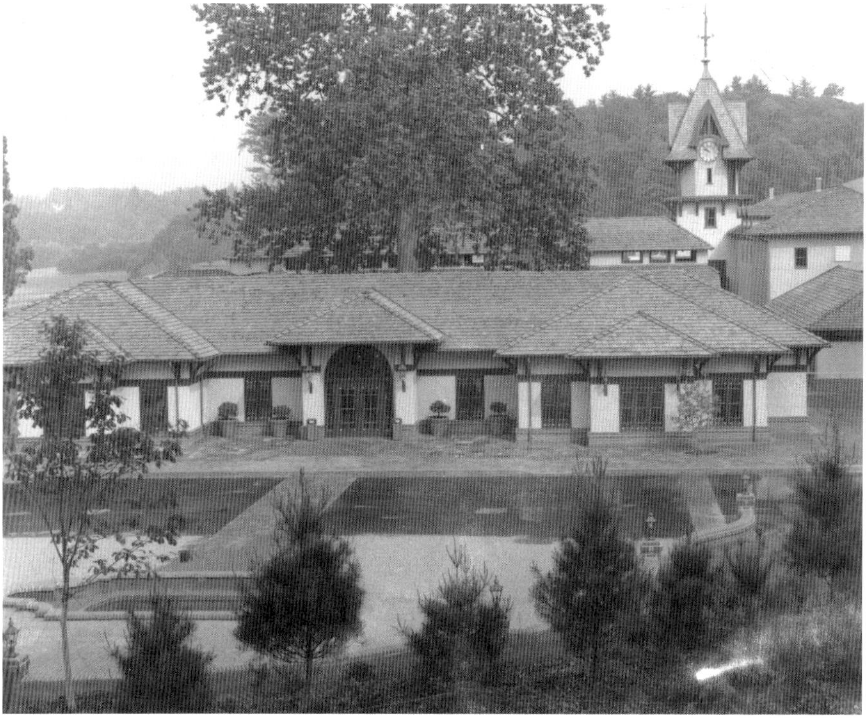

Biltmore Estate Winery, Asheville, North Carolina. Biltmore Estate Winery, housed on Biltmore Estate in a renovated dairy complex designed by renowned architect Richard Morris Hunt, opened its doors to the public in 1985. Each year more than 500,000 tour the winery, making it the most visited winery in the United States. *Courtesy of Biltmore Estate Winery.*

Despite its long historical affinity with grape growing and winemaking, North Carolina grape growers generally organized at the local level or were part of various state and local agricultural societies. During the 1970s, that situation changed. Growers with ties to the state's two great wine traditions organized. The first to organize were the muscadine grape growers, who formed what is now the North Carolina Muscadine Grape Growers Association. The Piedmont Grape Growers Association, the first organization for *vinifera* producers, was the seed for today's North Carolina Vinegrowers Association. These and other groups support education, marketing and research.

In addition to Duplin Winery, there are a number of other muscadine wineries in North Carolina. In 2010, among the members of the North Carolina Muscadine Association were the following wineries: Adams

Vineyards, Willow Spring; A Secret Garden Winery, Pikeville; Bannerman Vineyard, Burgaw; Benjamin Vineyards & Winery, Graham; Bennett Vineyards, Edward; Chateau Laurinda, Sparta; Cypress Bend Vineyards, Wagram; Dennis Vineyards, Albemarle; Garden Gate Vineyards, Mocksville; Ginger Creek Vineyards, Taylorsville; Glen Marie Vineyards & Winery, Burlington; Grapeful Sisters Winery, Tabor City; Hinnant Family Vineyards, Pine Level; Lu Mil Winery; Martin Vineyards, Knotts Island; Moonrise Bay Vineyard, Knotts Island; Old Stone Vineyard & Winery, Salisbury; Rocky Road, Midland; Woodmill Winery, Vale; and Uwharrie Vineyards, Albemarle. Many of these are family establishments. These wineries are not only committed to making good wine but also to improving life in their communities. Located primarily in the Piedmont and sandhills/coastal regions, these wineries offer a wide variety of wines, ranging from Noble, scuppernong and Carlos to strawberry, blueberry, peach, syrah, chardonnay and Chambourcin.

In 1985, there was bad news for North Carolina wineries. As a result of a Supreme Court ruling, the State of North Carolina ended its "preferential tax rate for native wines." In 1984, the United States Supreme Court ruled in the case of *Bacchus Imports, Ltd. v. Dias* that Hawaiian legislation that provided a tax exemption for locally produced brandy (Okolehao) and fruit wines violated the Commerce Clause of the United States Constitution, as it "discriminated in favor of local products." In an effort to ameliorate the damage to the wine business, the North Carolina Grape Growers Association agitated for other state assistance.

As a result, in 1986, the State of North Carolina founded the North Carolina Grape Growers Council. Demonstrating the wisdom of that move, in 1987, the North Carolina Grape Growers Council successfully lobbied for legislation to direct state excise tax monies collected on wine into research and promotion of North Carolina's valuable wine industry. In 2005, the council moved from the North Carolina Department of Agriculture to the North Carolina Department of Commerce. As a result, the Secretary of Commerce appoints the eleven members of the council. In 2006, reflecting the council's expanded role in research, education and tourism, the North Carolina Grape Growers Council became the North Carolina Wine & Grape Council. Consequently, the North Carolina

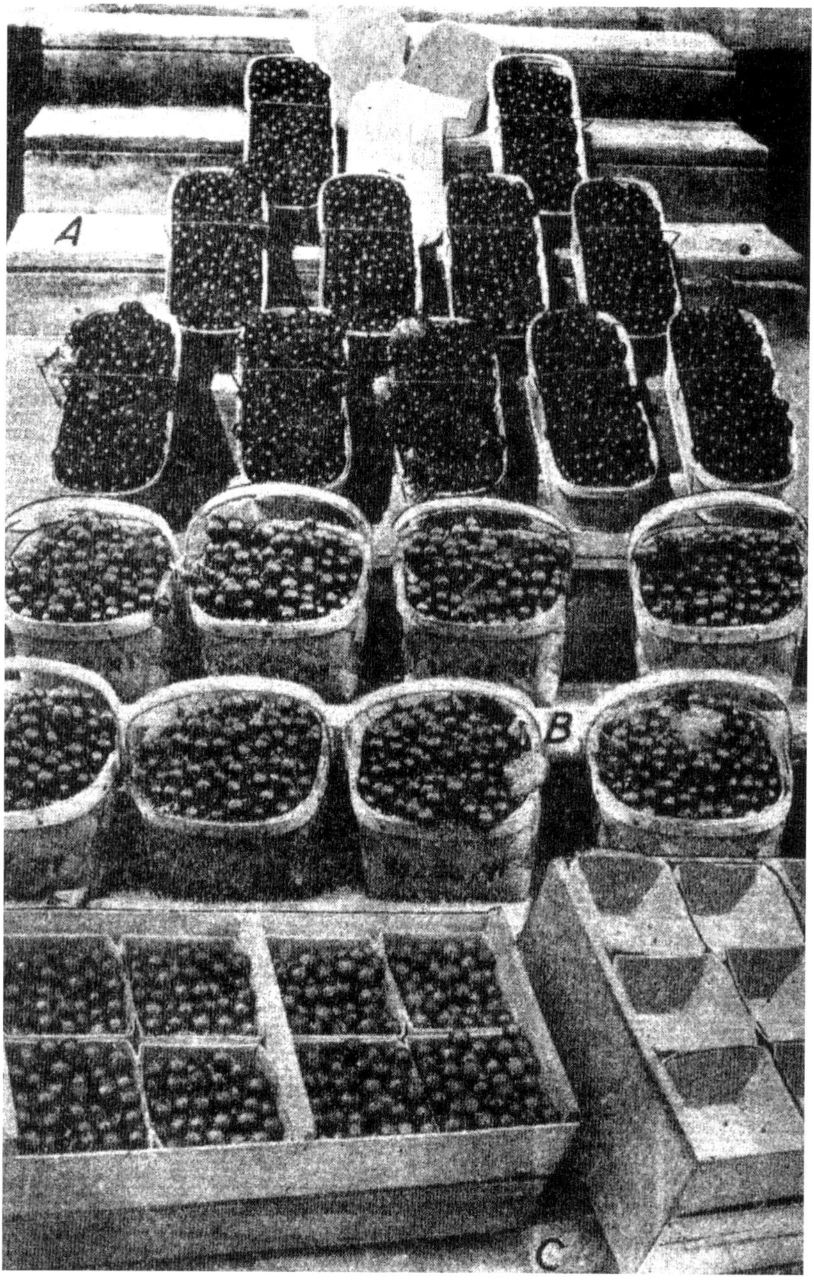

Handpicked muscadine grapes ready for shipment, circa 1916. According to the U.S. Department of Agriculture, in the early twentieth century, muscadine grapes were generally used for wine production. In 1916, the largest muscadine wineries in the South were in Norfolk, Virginia, and Wilmington, North Carolina—Garrett & Company and Sol Bear & Company, respectively. *George C. Husman,* Muscadine Grapes, *courtesy of the author.*

Wine & Grape Council sponsors research to improve grape varieties and the production and storage concerns for grape growers. In addition, the council provides education opportunities and marketing services for North Carolina's winemakers and grape growers.

The number of North Carolina wineries has grown steadily, and the acreage devoted to *vinifera* and French-American hybrids, especially in the Piedmont and western counties, has increased. Muscadine production continues its historic domination of eastern North Carolina. So from the mountains to the sand hills, North Carolina is a great wine-producing state.

Epilogue

Present and Future

We've got a great future.
—David G. Fussell Sr.

In 2010, North Carolina had ninety wineries, large and small, great places to visit and sample the rich wine experience of North Carolina. Nationally, North Carolina ranked eighth in wine production and ninth in grape production. Among North Carolina's larger wineries are Duplin, Biltmore Estate Winery and Shelton Vineyards.

Shelton Vineyards

Shelton Vineyards, founded in 1999 by Charles M. and Ed Shelton, is the "largest family-owned estate winery in North Carolina." The Sheltons planted more than 160 acres of vineyard on 265 acres that was once dairy land. The thirty-three-thousand-square-foot winery opened in 2000 and celebrated its tenth anniversary in 2010. Located near Dobson, Surry County, North Carolina, in the Yadkin Valley, the vineyard's beautifully landscaped grounds and award-winning wines characterize the best of North Carolina estate wineries. For example, the judges awarded medals

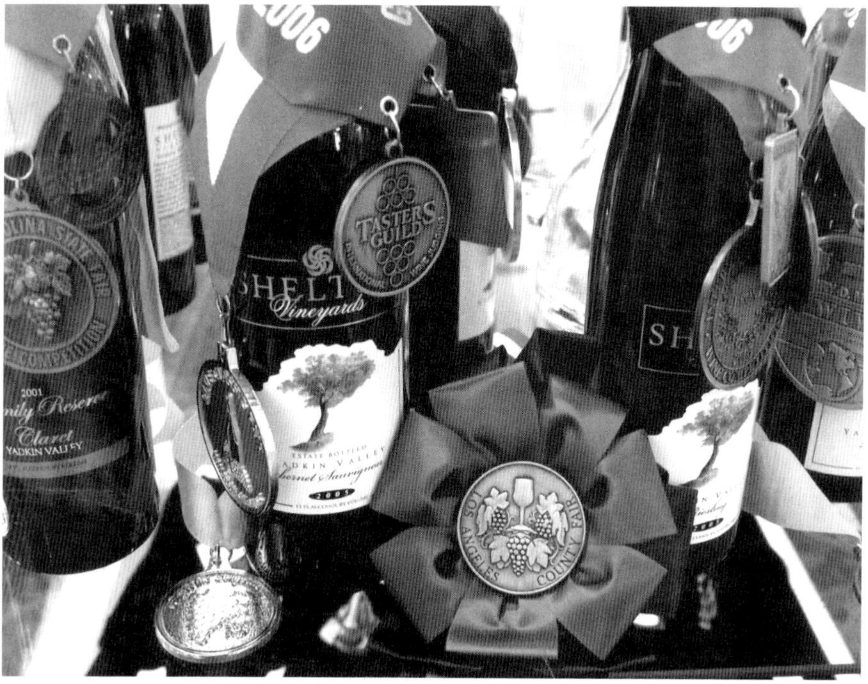

Shelton Vineyard awards. Throughout its history, North Carolina has produced many award-winning wines. Shelton Vineyards, home of many award-winning wines, celebrated its tenth anniversary in 2010. *Courtesy of George A. Denka.*

to each of the six Shelton Vineyard wines entered in the 2010 Finger Lakes International Wine Competition.

In addition to their pursuit of excellence in their winemaking, the Sheltons have supported viticultural education in the region. Shelton Vineyard also joined with other Yadkin Valley wineries in seeking American Viticultural Area status for the Yadkin Valley. As co-founder Charles Shelton noted, the Yadkin Valley is well placed for wine production, as there are two and a half million people within a 250-mile radius. So the market is there for the right product. Since 2003, George A. Denka, who joined Shelton in 2001, has served as president of Shelton Vineyards.

Other North Carolina Wineries

In 1991, winemaking returned to Polk County, home of the Tryon grapes, in western North Carolina. Lee Griffin and Marsha Cassedy, his wife, of Rockhouse Vineyards and Winery, planted their first vines. Rockhouse, bonded by the State of North Carolina in 1998, is a small family winery featuring *vinifera* wines and mountain vistas. Located near Tryon, the winery's vineyard covers ten acres. Other vineyards and wineries, such as Green Creek Winery in Columbus, followed their lead. Green Creek, another favorite with picnickers, opened in 2005.

North Carolina is blessed with many great wineries. The following sample from the Yadkin Valley introduces a few of the North Carolina wineries of particular interest and hints at the many special stories these wineries and their owners can tell. Among these Yadkin Valley wineries are RagApple Lassie, RayLen, Hanover Park, Stony Knoll, Round Peak and the Old North State Winery & Brewery. Old North State has special

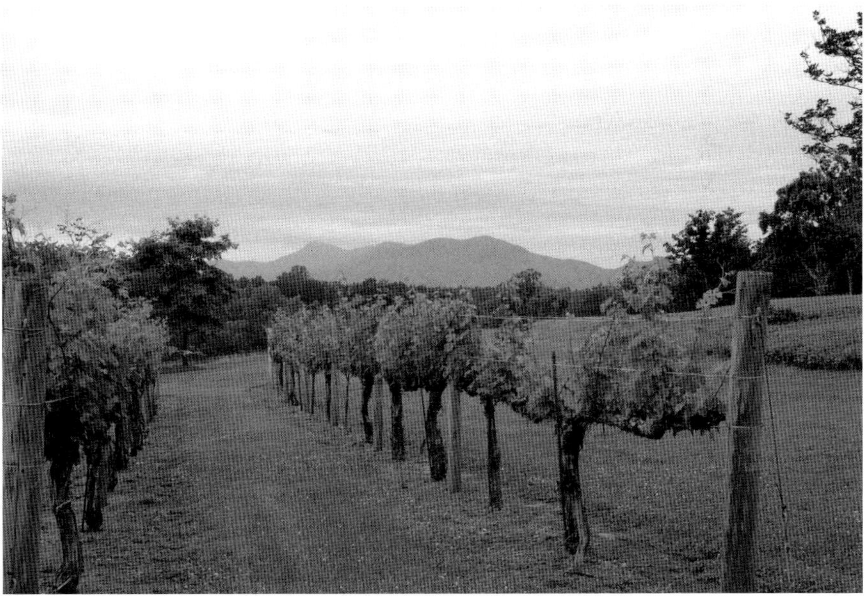

Rockhouse Vineyards, Tryon. Lee Griffin and Marsha Cassedy planted their first vines in 1991. In 1998, Rockhouse became a bonded North Carolina winery. In 2010, Rockhouse Vineyards had ten acres of producing vines. *Photograph by Jacob Helsley, courtesy of Lee Griffin.*

historical significance, as it was the first cooperative winery in North Carolina. In 2002, thirty-eight vineyard owners founded the winery with a Golden Leaf grant from the State of North Carolina. A partnership now operates the winery in Mount Airy. Hanover Park in Yadkinville, on the other hand, has a romantic story. Two teachers married and, during their honeymoon in the south of France, fell in love with French wine. Returning home, they followed their dream and started a winery in 1996.

RayLen Vineyards & Winery, in Mocksville, is also the result of a European trip. A trip to Italy inspired owners Joe and Joyce Neely to open a winery. RayLen produced its first wine in 2000 and became a bonded winery in 2001. At that time, there were only four other Yadkin Valley wineries. In 2005, the Mid-Atlantic Wine Festival named RayLen's "Category 5" wine as the "Best in Show." RagApple Lassie Vineyards & Winery in Boonville has a different story. Frank W. Hobson Jr. is a third-generation farmer. Faced with decreasing tobacco allotments, Hobson looked for an alternative crop. Several visits to Shelton Vineyards convinced him to explore grapes. Soon the idea of growing grapes expanded into RagApple Lassie Vineyards & Winery. The Hobsons named their winery for Hobson's 4-H Grand Champion Show calf. Stony Knoll Vineyards, in Dobson near Rockford, North Carolina, sits on another family farm. The Coe family has owned the farm since 1896. Travel was also a factor in the creation of Round Peak Vineyards in Mount Airy. Founders George and Susan Little and Janet and Lee Martin visited Napa Valley in 1998. They purchased land for the vineyard in 1999 and planted their first grapes in 2000.

In 2001, the State of North Carolina recognized the state's long and historic love affair with the Great White Grape. The General Assembly named the scuppernong the official state fruit. Representative T. LaFontine "Fountain" Odom of Charlotte championed the effort. Essays about the Mother Vine written by fourth graders from Manteo may have swayed the legislature's decision. Apparently, the strawberry was the other principal contender for state fruit. North Carolina and the scuppernong have a long heritage, as seen in North Carolina's official state toast, adopted in 1957. The official toast includes the line "where the Scuppernong perfumes the breeze at night." In 2007, North Carolina's grape and wine industry was worth $813 million and ranked tenth in the United States.

On February 7, 2003, the United States Bureau of Alcohol, Tobacco, Firearms and Explosives approved the Yadkin Valley AVA (American Viticultural Area). In 2008, there were 190 American Viticultural Areas in the United States. An AVA label indicates that 85 percent of the grapes in a particular vintage came from a designated geographical location. The Yadkin Valley AVA, located in the Piedmont and foothills, was North Carolina's first. The appellation authorized the wineries to bottle wines with the Yadkin Valley designation. By 2005, there were twenty-three wine producers in the region. In 2010, the Yadkin Valley AVA includes over one million acres in Surry, Yadkin, Wilkins, Davie, Davidson and Forsyth Counties and wineries such as Shelton Vineyards in Surry County and Westbend Vineyards in Forsyth County. Surry Community College in Dobson was an important part of the AVA. The college offered degree and non-degree programs in viticulture and enology (the study of wine and of the making of wine). In 2005, Surry partnered with Davidson County Community College to offer enological and viticultural training in Davidson and Davie Counties.

On May 27, 2008, the Swan Creek AVA became North Carolina's second area. This appellation recognized the commitment of the area to grapes, vineyards, wineries and related enterprises associated with wine tourism. The Swan Creek AVA covers ninety-six thousand acres in Wilkes, Yadkin and Iredell Counties. The wineries of Swan Creek include Raffaldini Vineyards in Ronda; and Laurel Gray Vineyards, Shadow Spring Vineyard, Buck Shoals Vineyards and Dobbins Creek Vineyards, all located in Hamptonville. Supporters of the Swan Creek AVA, such as Raffaldini Vineyards, cite the special microclimate of the area and the rarely available French and Italian wines of the region.

North Carolina's third federally recognized AVA district is the Haw River Valley. The Haw River Valley AVA, recognized in 2009, covers 868 square miles in Alamance, Caswell, Chatham, Guilford, Orange and Rockingham Counties. Member wineries include Grove Winery & Vineyards, Gibsonville; Benjamin Vineyards, Graham; Glen Marie Vineyards & Winery, Burlington; Silk Hope Winery, Pittsboro; and Wolfe Wines, Snow Camp. Also, in 2007, Polk County vintners and grape growers discussed applying for an AVA designation for the Tryon-foothills wine-producing area. To many supporters, attaining

AVA status is a major step toward improving the image of North Carolina wines.

In addition, North Carolina boasts numerous wine trails. The state's wine trails include, but are not limited to: the Sandhills Wine Trail, the Duplin/I-40 Wine Trail, the Uwharrie Mountains Wine Trail, the Piedmont Heritage Wine Trail, the Mountain Lakes Wine Trail, the Haw River Wine Trail and several Yadkin Valley trails. Shorter or mini-trails include, among others, the Brunswick Wine Mini-Trail and the Cherokee Wine Mini-Trail.

In 2009, North Carolina boasted, besides its wineries and wine producers, over 425 grape growers. Of the more than two thousand acres of vineyards planted in North Carolina, 65 percent produce *vinifera* grapes, 29 percent produce *rotundifolia*, 4 percent produce European/American hybrids and 2 percent produce *labrusca*. North Carolina grape growers harvested 4,800 tons of grapes in 2009.

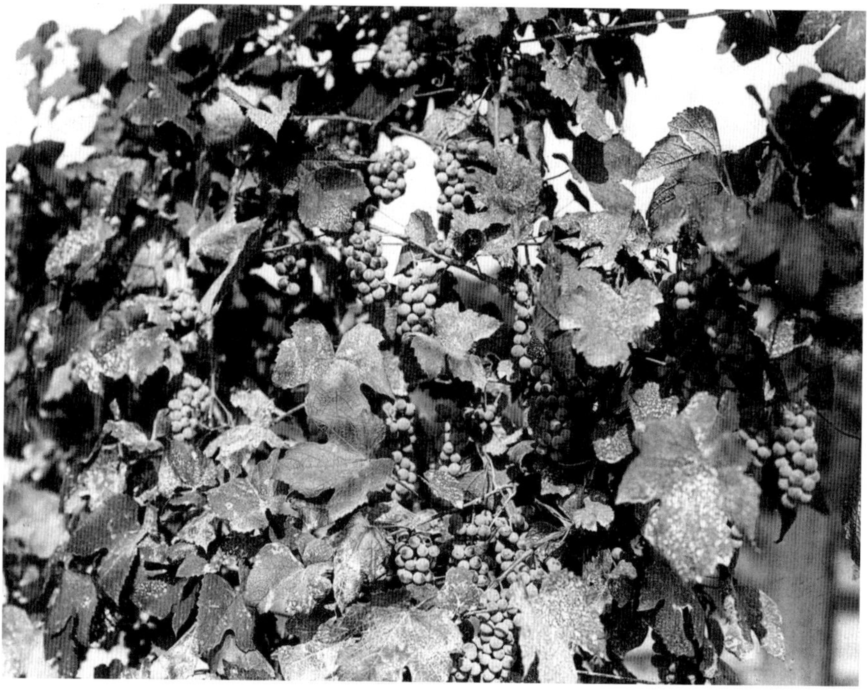

Niagara Grapes, Moore County, North Carolina. The Niagara grape is a cultivar of *Vitis labrusca* and is used in white wine production. *Courtesy of the North Carolina State Archives.*

The future of North Carolina wines is one of diversity and challenge. For example, in 2006, Raffaldini Vineyards on Brushy Mountain opened, featuring Italian wines, and wine is making inroads on the NASCAR circuit. Richard Childress, owner of three Nextel Cup teams, is creating fine wine at Childress Vineyards in Davidson County. Childress connects with racing fans through his Fast Track Wine Club and other marketing ventures. For example, in 2005, Childress Vineyards debuted Fine Swine Wine in honor of the annual Barbeque Festival in Lexington. Diversity is also essential for the survival and continued growth of North Carolina's wine duality: *vinifera* v. *rotundifolia.* Health claims, byproducts (jam, jelly, syrup), dermatological uses, concerts, tastings, hayrides, restaurants and tourism all have a role in the future of the North Carolina wine industry. The potential health benefits of resveratrol may play a major role. In addition to identified heart benefit, some scientists also see resveratrol as a factor that contributes to human longevity.

The challenges, current and potential, are many: educating consumers on the quality and vintages of North Carolina wines, developing region-specific viticultural research, improving viticulture and enology education, creating a consumer identify for North Carolina wines, improving quality control in the vineyards and wineries, developing active producers' organizations, maintaining the distinctiveness of North Carolina wines, expanding the market share of North Carolina wines nationally and internationally, coordinating a cohesive marketing strategy, developing strategies for climatic changes in temperature and rainfall and addressing the viability of the current North Carolina winery bubble.

On the plus side, however, North Carolina has taken a major step forward in viticultural education. The new Shelton-Badgett Center for Viticulture and Enology at Surry Community College is poised to make North Carolina a regional viticultural education center. Charles Shelton of Shelton Vineyards was a major factor in the development of the center. Surry Community College, located in Dobson, North Carolina, offers one of three accredited courses in viticulture and enology on the East Coast. Cornell University in Ithaca, New York, and Virginia Tech (Virginia Polytechnic Institute and State University) in Blacksburg, Virginia, offer the other accredited East Coast programs. Surry Community College also has a working winery, so in addition to classroom learning, its students receive firsthand job experience.

North Carolina has the tools, the wineries and the wines to claim its rightful place in American viticulture. For example, at the first Budbreak Wine Festival held in 2010 at Old North State Winery, several of the judges commented on the quality of North Carolina wines. Pierre Louis Teissedre, professor of enology at the University of Bordeaux, France, commented on the "wonderful variety of [North Carolina] wines." To reach its potential, North Carolina needs to address pressing issues, solidify its gains and focus on a united future for all its grape growers and wine producers. In addition, the state of North Carolina faces the unique challenge of claiming and celebrating a grape heritage that stretches back more than four hundred years, while affirming its remarkable *vinifera* success. *Viva la difference!*

Glossary of Wine Terms

American Viticultural Area (AVA): a special designation of a particular winemaking area. The United States Bureau of Alcohol, Tobacco, Firearms and Explosives must approve the AVA appellation, which allows winemakers to bottle wine with a label indicating that the wine came from that region.

appellation: a distinguishing name, in this case, for a wine-producing area

Catawba: a variety of the native fox grape *Vitis labrusca*

cave: a chamber, often underground, where wine is stored in kegs and aged

cultivar: a variety of plant developed or intentionally selected and maintained for cultivation

enology: the art and science of winemaking

generic wine: an American wine named for a particular region in Europe; for example, Chianti or Chablis

horticulturalist: one who studies the science of cultivating plants

Mother Vine: an ancient scuppernong vine on Roanoke Island; perhaps the oldest cultivated vine in the United States

muscadine: the popular name for *Vitis rotundifolia* grapes

pomologist: a scientist or other individual who studies the cultivation of fruit trees

PROPRIETARY WINE: a wine with a winery's trademark name; for example, Shelton Family Reserve Claret

SCAFFOLDING: a wooden framework to support scuppernong and other grapevines

SCUPPERNONG: a variety of *Vitis rotundifolia*; also known as "the great white grape"; the first identified native North Carolina grape

VARIETALS: a word that indicates the grapes from which a wine was made. By law, a varietal wine must contain at least 75 percent of the labeled wine.

VINEYARD: a planned planting of grapes

VINTAGE: date on which grapes are harvested

VINTNER: a winemaker

VITICULTURE: the study of grapes and grape growing

VITIS LABRUSCA: the species name of the fox grape, one of America's native grape varieties; for example, the Concord grape

VITIS ROTUNDIFOLIA: species name for native American grapes known as muscadines. Scuppernong is a type of muscadine grape.

VITIS VINIFERA: species name for European grape cultivars such as chardonnay or cabernet sauvignon

For Additional Reading

Adams, Leon D. *The Wines of America*. New York: 1985. Third Edition.

Cathey, C.O. "Sidney Weller, Ante-Bellum Promoter of Agricultural Reform." *North Carolina Historical Review* 21 (1954): 1–17.

Drew, Randy. *The North Carolina Muscadine: A Historical Timeline.* Wilmington, NC: n.d.

Gohdes, Clarence. *Scuppernong: North Carolina's Grape and Its Wines.* Durham, NC: 1982.

Green, Wharton Jackson. *Tokay Vineyard, near Fayetteville, N.C. with Essay on Grape-Culture by the Proprietor.* Boston: 1883.

Heath, Mortimer O. *Sketches in North Carolina USA 1872 to 1878: Vineyard Scenes*. Edited by H.G. Jones. Raleigh, NC: 2001.

Mills, Joseph, and Danielle Tarmey. *A Guide to North Carolina's Wineries.* Winston-Salem, NC: 2007.

Pinney, Thomas. *A History of Wine in America: From the Beginnings to Prohibition.* Berkeley, CA: 1989.

———. *A History of Wine in America from Prohibition to the Present.* Berkeley, CA: 2005.

Watson, Pamela. *Carolina Wine Country: The Complete Guide.* Greenville, NC: 1999.

Index

C

F

G

H

M

About the Author

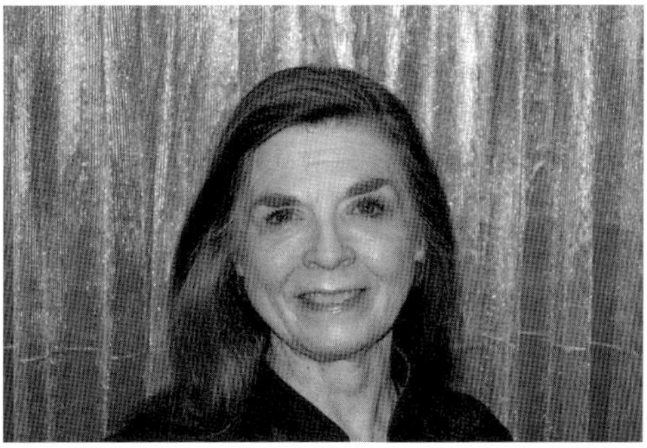

Alexia Jones Helsley is an archivist and historian with deep roots in western North Carolina. Her parents live in Hendersonville, and her father, Dr. George A. Jones, is a native of Saluda, North Carolina. She is a founding member and program vice-president of the Henderson County Genealogical and Historical Society and has published widely on the history of North and South Carolina. Her North Carolina titles include *A Guide to Historic Henderson County, North Carolina* and the *Henderson County (N.C.) Revolutionary Pensioners of 1840*, 1997 recipient of the Willie Parker Peace Prize (North Carolina Society of Historians). Other relevant research interests include the Battle of Kings Mountain, Mountain Page Baptist Church, Carolina migration trails and the Pace family of western North Carolina. Her Pace ancestors moved from Virginia into eastern North Carolina and eventually settled near Saluda. Helsley, an instructor in history for the University of South Carolina–Aiken, currently serves as president of the Pace Society of America. In addition, she is a member of the Old Exchange Commission, and in 2006, the South Carolina State

Historic Records Advisory Board presented her with the Governor's Archives Award.

While researching the history of western North Carolina, she became interested in North Carolina's grapes and the state's long tradition of winemaking.